Exercises for Weather and Climate

FOURTH EDITION

Greg Carbone

University of South Carolina—Columbia

Prentice Hall

Upper Saddle River, NJ 07458

Executive Editor: *Dan Kaveney*
Assistant Editor: *Amanda Griffith*
Media Editor: *Michael Banino*
Special Projects Manager: *Barbara A. Murray*
Manufacturing Manager: *Trudy Pisciotti*
Cover Design: *Joseph Sengotta*
Cover Photo: *Fog Over East River Mountain/Melvin L. Grubb, Grubb Photo Service, Inc.*

© 2001 by Prentice-Hall, Inc.
Upper Saddle River, New Jersey 07458

Printed in the United States of America
10 9 8 7 6 5 4 3 2

ISBN 0-13-016798-3

Prentice-Hall International (UK) Limited, *London*
Prentice-Hall of Australia Pty. Limited, *Sydney*
Prentice-Hall Canada Inc., *Toronto*
Prentice-Hall Hispanoamericana, S.A., *Mexico*
Prentice-Hall of India Private Limited, *New Delhi*
Prentice-Hall of Japan, Inc., *Tokyo*
Pearson Education Asia Pte. Ltd., *Singapore*
Editora Prentice-Hall do Brasil, Ltda., *Rio de Janeiro*

TABLE OF CONTENTS

PREFACE

This lab manual contains exercises for an introductory college-level meteorology course. It combines data analysis, problem solving, and experimentation, with questions designed to encourage critical thinking. The review questions at the end of each chapter are meant to measure comprehension, but more importantly to extend students' thinking about atmospheric processes.

The manual was written to complement either Aguado and Burt's *Understanding Weather and Climate, 2d edition* or Lutgens and Tarbuck's *The Atmosphere, 8th edition*. It shares organization, concepts, and graphics with both. However, it should be appropriate for any introductory meteorology laboratory course. I hope that the exercises provide students with a chance to apply what they've learned in lectures or their text. While nearly all questions can be answered from material contained within the manual, a few in each lab will demand that students consider lecture notes or material from their introductory text.

The basic structure of the manual remains the same as previous editions, but revisions have been made to nearly every lab. Labs 10, 11, and 14 have expanded sections on weather map analysis, mid-latitude cyclones, and climatic variability and change respectively. Many new graphics appear in this edition and a majority of old ones have been redrafted. The inclusion of 6 new interactive computer modules marks the most significant change. These modules complement Labs 2, 4, 6, 13, and 14. Questions at the end of each of these labs introduce students to the software, but they are encouraged to ask their own questions and play. As with previous editions, GeoClock—Joseph Ahlgren's shareware displaying geographical and seasonal variations in sunlight—and an energy balance model written by James E. Burt accompany the manual. Lab 3 requires GeoClock; Labs 5 and 18 require the energy balance model. The CD enclosed at the back of the manual contains all software.

I thank those who contributed to this edition. Mark Anderson, Cecil Keen, Scott Kirsch, John Knox, Scott Robeson, Robert Rohli, Steven Silberberg, and Anthony Vega reviewed the last edition and offered very helpful suggestions. Eric Stevens drafted many of the new figures. Robert Ellis, Xin Kang, and Haiyun Yang assisted with the computer modules. Jim Burt, Bill Kiechle, Cary Mock, Helen Power, Jennifer Rainman, and Donald Yow provided advice, data, and graphics. Mel Grubb took the cover photo and Carol Hall found it for me. Teaching assistants and students from an introductory meteorology course at the University of South Carolina provided helpful critique. Michael Banino, Amanda Griffith, Christine Henry, Dan Kaveney and others at Prentice Hall contributed their expertise throughout the project. Finally, I am indebted to Karen Beidel who gave her time generously, offering extraordinary talents to every aspect of the manual's written and digital form. Without her, there would be no lab manual.

SOFTWARE NOTES

The accompanying CD contains three folders—GeoClock, EBM, and mods4car—that should be copied to your hard disk. Be sure to maintain the folder structure found on the CD!

GeoClock

GeoClock for Windows will run on a minimal Windows system—VGA, 4 MB RAM, 80286 processor. However, the recommended minimum requirements are: 486DX33 processor, 8 MB RAM, 256 color display, 800x600 display. The CD contains a GeoClock folder. You can install this shareware by simply dragging this folder to a desired place on your hard drive. Start the software by clicking on the GEOCKWIN.EXE (Windows) icon. When you first use GeoClock, you will be prompted to enter setup information including: home town name, latitude, longitude, and time zone (there are facilities to acquire this information).

GeoClock is marketed as shareware. It can be used free as part of Lab 3, but if further use is desired, a more complete version should be purchased via the MAILER file, the Help | Register menu, or the web (http://home.att..net/~geoclock/).

Energy Balance Model

The energy balance model runs under DOS. You can install this software by dragging the EBM folder to an appropriate place on your hard drive. To run the software using any version of windows, double-click the program item (ebmodel.exe) or highlight the program and press the "enter" key. When you are finished with the program type "exit" in the command box.

Interactive Modules

The interactive computer modules can be accessed on any platform with a web browser that supports JavaScript and Java 1.1. Install the modules by dragging the "mods4car"folder to your hard drive, **preserving the directory structure.** The "mods4car" folder contains individual folders for each module and an "index.html" file. Open your browser, use the "open file" or "open page" command to select the "index.html" file, and choose the appropriate module.

Vertical Structure of the Atmosphere

Materials Needed
- calculator
- ruler

Introduction
Our first lab introduces the concept of atmospheric pressure. We will construct and interpret a number of graphs to measure how pressure, density, and temperature change with height above the earth's surface. We will focus on how these relationships are expressed in the troposphere, which is where most weather processes occur.

Changes in Atmospheric Pressure with Height
The atmosphere is a compressible fluid, made up of gases whose molecules are pulled to the earth's surface by gravity. As a result, the molecules that make up the atmosphere are most compressed close to the earth's surface and atmospheric density decreases most rapidly with height there (Figure 1-1).

While the boundary between the earth's surface and the atmosphere is obvious, there is no clear "top" to the atmosphere. It thins out with increasing height, but never actually ends. (The phenomenon is analogous to repeatedly dividing numbers in half. Each division produces a smaller number, but theoretically one never reaches zero.) However, since very few gas molecules within earth's gravitational field exist beyond 100 km, we can consider this height an arbitrary "top" of the atmosphere.

We may use a simple rule to describe the rate at which density decreases with height: no matter where you begin in the atmosphere, if you ascend 5.6 kilometers (km), you will cut in half the atmospheric mass above you when you started.

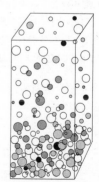

Figure 1-1

1

1. Using the above rule, indicate the percent of the atmosphere above and below each height in Table 1-1.

TABLE **1-1**

Height (km)	% of Atmosphere Above	% of Atmosphere Below
33.6		
28.0		
22.4		
16.8		
11.2		
5.6		
Sea Level	100%	0%

2. Use the data in Table 1-1 to construct a graph on the next page. The vertical axis should be divided into 17 equally-spaced intervals. Label this axis "Height Above the Surface (km)" and label the intervals. Label the horizontal axis "Percent of the Atmosphere Above" and label its intervals (0–100%). You will use the graph to answer questions 3–7.

3. A jet airplane travels at about 11.2 km above sea level (~37,000 ft.). Approximately what fraction of the atmosphere is <u>above</u> these jets?_____ %

4. Approximately what fraction is <u>below</u> the summit of Mt. McKinley (Denali) in Alaska (6.19 km, 20,320 ft.)? _____ %

5. Approximately what fraction is <u>above</u> Pike's Peak (4.34 km, 14,110 ft.)? _____ %

Since barometric pressure reflects the weight of the atmosphere above a point, there is also a close relationship between height and atmospheric pressure. We can assume that one hundred percent of the atmospheric mass lies above sea level and exerts a pressure of approximately 1000 mb. Since atmospheric mass at 5.6 km is 50% of its sea-level value, the pressure at this height is half of that exerted at sea level.

6. *Using this relationship, add a pressure scale beneath the percent scale in your graph and label it "Pressure (mb)."*

7. *Using this new scale, estimate the pressure at the three levels used above.*
    Height of jet airplanes _____ *mb*

    The top of Mt. McKinley (Denali) _____ *mb*

    The top of Pike's Peak _____ *mb*

Changes in Temperature with Height

We define four layers of the atmosphere (the *troposphere, stratosphere, mesosphere,* and *thermosphere)* according to their average *lapse rate*—the rate at which temperature changes with height (Figure 1-2). While the lapse rate at any given time or place will differ from this average, the figure provides a starting point for understanding the temperature profile of the atmosphere. The *tropopause, stratopause,* and *mesopause* mark the top (end) of each layer.

8. *What does the lapse rate in the troposphere suggest about where this layer gets its heat?*

9. *Ozone is a good absorber of ultraviolet radiation from the sun. How does its highest concentration at 20–30 km influence the stratospheric lapse rate?*

4

11. a. Which station has the higher surface temperature?
    b. Which station has the higher temperature at 10 kilometers?
    c. At 16 kilometers?

12. How does the temperature profile in the lowest 1 kilometer differ between the two stations?

13. The tropopause marks the top of the troposphere and is defined as the height of coolest tropospheric temperature. Record the tropopause height at Key West and Fairbanks and their respective temperatures.

	Tropopause Height	Tropopause Temperature
Key West	10000 m	-43°C
Fairbanks	10000 m	-51°C

14. How would you characterize the relationship between average temperature and the thickness of the troposphere?

As your graph shows, there are some situations where temperature increases with height in certain layers of the troposphere. Since this is opposite the norm, such layers are called *inversion* layers.

15. Circle a layer of either temperature profile where an inversion occurs.

Review Questions

In your own words, describe how air pressure, density, and temperature vary with height in the troposphere.

The thickness of the troposphere varies from place to place and from day to day. What influences this thickness?

Do you think air pressure varies faster horizontally or vertically? (As a reference, an average hurricane has a radius of approximately 600 km and a central pressure of 950 mb.)

EARTH SUN GEOMETRY

Materials Needed
- calculator (with trigonometric functions)
- ruler
- drawing compass

Introduction
What causes the seasons and changes in the amount of daylight we receive? The diagrams and exercises in this lab show how earth–sun geometry influences these variables. We examine the earth–sun relationship early in our study of weather and climate because most atmospheric processes are ultimately driven by spatial variations in solar energy.

Earth–Sun Relationships
The distance between the earth and the sun averages about 150 million kilometers (93 million miles). Because of this distance and the earth's relatively small size compared to the sun, it is reasonable to assume that the sun's rays strike the nearly spherical earth in straight paths.

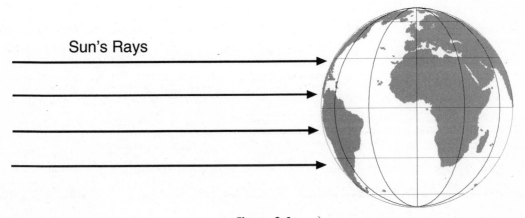

Sun's Rays

Figure 2-1

The earth's axis of rotation is tilted 23½ degrees from the perpendicular to the *plane of the ecliptic*—the plane on which the earth revolves around the sun. This tilt is oriented in the same direction throughout the year with the North Pole presently pointing toward the North Star, Polaris. Figure 2-2 (not to scale) shows that the Northern Hemisphere is tilted toward the sun during its summer months and away from the sun during its winter months.

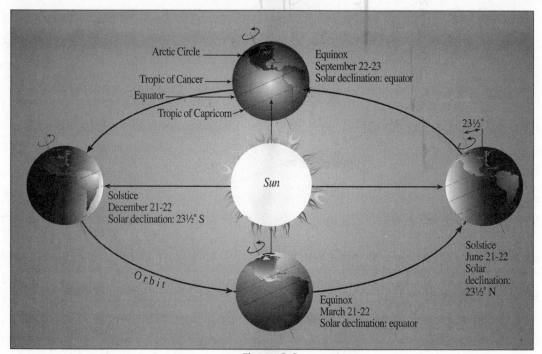

Figure 2-2

Our seasons occur because of this tilt. As the earth revolves around the sun, the sun's direct rays strike different latitudes. When the Northern Hemisphere is tilted toward the sun, it receives the more direct and, therefore, more intense rays of the sun. Locations in the Southern Hemisphere receive less direct solar radiation. Six months later, when the Southern Hemisphere is tilted toward the sun, it receives the more direct solar radiation.

Figure 2-3 shows how the sun's rays strike the earth on December 22, the Northern Hemisphere's winter solstice. At solar noon on this date, the sun's rays are perpendicular to the earth's surface at 23½° S (location A). As we move away from 23½° S, we see that the rays of the sun strike the earth's surface at progressively lower angles. Location B is at the equator, location C is at 30° N latitude, and location D is at 66½° N latitude.

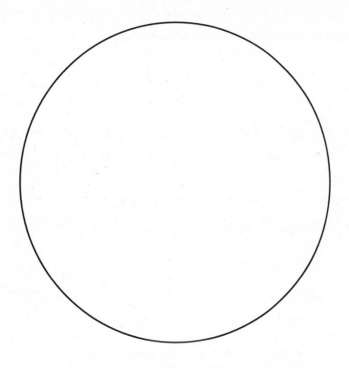

Sun's rays striking the earth on June 21

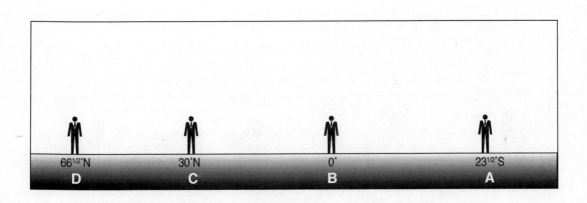

66½°N   30°N   0°   23½°S

D   C   B   A

Profile view at the earth's surface: Solar noon, June 21

Your diagrams should show that sun angle varies with season and location. Since such variability greatly influences weather patterns, it is useful to be able to calculate the noon sun angle for a given latitude. We must first define a few terms:

- *solar declination*—the latitude at which the sun is directly overhead at solar noon
- *zenith angle*—the angle between a point directly overhead and the sun at solar noon
- *solar elevation (sun) angle*—the angle of the sun above the horizon at solar noon

To calculate the noon zenith angle (A), simply find the number of degrees of latitude separating the location receiving the direct vertical rays of the sun and the location in question. The sun angle (B) is calculated by subtracting the zenith angle from 90 degrees. Figure 2-5 shows how this calculation is made at 40° N on December 22nd (the winter solstice), when the sun's rays are directly striking locations at 23½° S.

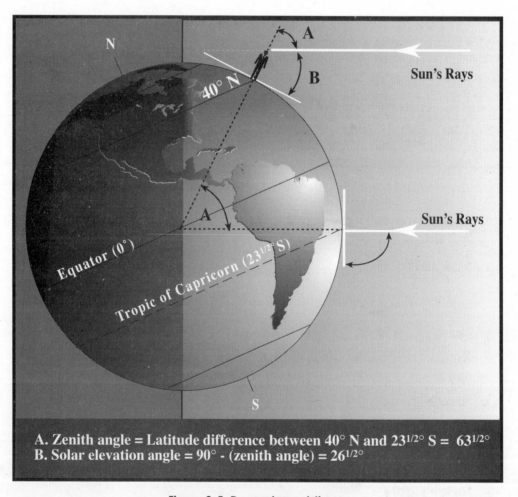

A. Zenith angle = Latitude difference between 40° N and 23^{1/2}° S = 63^{1/2}°
B. Solar elevation angle = 90° - (zenith angle) = 26^{1/2}°

Figure 2-5. December solstice.

Figure 2-2 gives the solar declination for the solstices and equinoxes, but what about the other days of the year? You can approximate the value of the solar declination using the following formula

$$\text{Solar declination} \approx 23.5 \bullet \sin (N)$$

where N = the number of days to the closest equinox, expressed in degrees. (By convention, N is positive between the March and September equinoxes and negative from the September to March equinoxes.)

For example, on April 20:

N = 30 (number of days from the closest equinox, March 21)
Declination $\approx 23.5 \bullet \sin (30°) = 23.5 \bullet (0.5) = 11.75°$ or 11° 45′ N

On December 9:

N = -78 (number of days from September 22,
    negative since it is between the September and March equinoxes)
Declination $\approx 23.5 \bullet \sin (-78°) = -22.90°$ or 22° 53′ S
                                    -22.99

2. **Calculate the solar declination on:**
    a. the March equinox _____   **b. the June solstice** _____
    c. your birthday _____   **d. today's date** _____

3. **Calculate the noon sun angle for New Orleans, USA (30° N) and for Helsinki, Finland (60° N) on each of the following dates:**

	New Orleans	Helsinki
a. March 21	_____	_____
b. June 21	_____	_____
c. September 22	_____	_____
d. December 22	_____	_____
e. Today's date	_____	_____

15

4. Use the following steps to measure the solar noon sun angle at your latitude:

    a. Set a pole with a known length at a right angle to a flat surface on the ground.

    b. Measure the length of its shadow.

    c. Divide the pole length by the shadow length to calculate the <u>tangent</u> of the sun angle θ.
        Tan θ = (Length of Pole) ÷ (Length of Shadow)

    d. Using a calculator or table of tangent values, convert the tangent ratio into an angle.
        θ = Tan⁻¹ [(Length of Pole) ÷ (Length of Shadow)]

        Today's sun angle: _____ °

5. Now calculate sun angle using the procedure in Figure 2-5. How does this value compare with the angle you measured? Can you account for any differences?

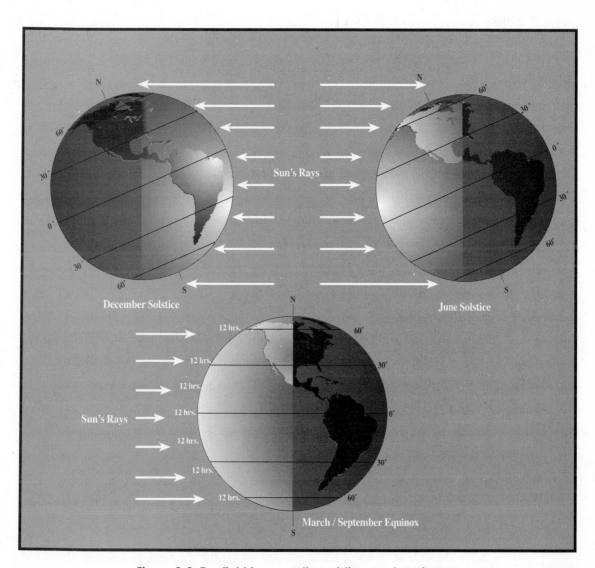

Figure 2-8. Daylight hours on the solstices and equinoxes.

Solar Radiation Receipt

Table 2-1 lists solar radiation received at solar noon at the top of the atmosphere for various northern hemisphere latitudes and dates. Values are in watts per square meter (Wm^{-2}).

TABLE 2-1

	Equator	23½°N	30° N	45° N	60° N	66½° N
March 21	1370	1256	1186	969	685	546
June 21	1256	1370	1361	1275	1101	1002
September 22	1370	1256	1186	969	685	546
December 22	1256	934	815	502	155	0

11. Calculate the difference in solar radiation received between 30° N and 60° N for each of the solstices and equinoxes.

Difference (Wm^{-2})   Difference (Wm^{-2})

March 21 _____   Sept. 22 _____

June 21 _____   Dec. 22 _____

12. Why is the seasonal range in solar radiation received greater at 60° N than at 30° N?

13. Why is the difference in solar receipt between 60° N and 30° N greater in the winter than in the summer?

14. *The values in Table 2-1 were derived from the following equation:*

$$I = I_o \cdot \sin \alpha$$

where: *I = solar intensity at the top of the atmosphere*
I_o = solar constant (1370 Wm^{-2})
α = solar angle

Calculate today's solar receipt at the top of the atmosphere at solar noon for your location.

Interactive Computer Exercise: Earth-Sun Geometry

This applet allows you to see how solar angle and solar intensity change with latitude and season. The questions below encourage you to compare solar intensity at one latitude against another. You might also want to experiment with the applet to answer your own questions.

Start the applet and note its initial settings:

Date:	*January 1*	Solar Declination:	*23.1° S*
Left diagram		Right diagram	
Latitude:	*0° (equator)*	Latitude:	*23.5° N*
Sun angle:	*66.9°*	Sun angle:	*43.4°*
Beam spreading:			
1 unit beam:	*1.087 surface units*	1 unit beam:	*1.455 surface units*

We could calculate the percentage difference in beam spreading between the two locations as:

$$\frac{1.455-1.087}{1.087} = 0.34$$

Solar intensity is 34% greater at the equator than at 23½° N on January 1.

Press the play button (▸) to simulate the earth's revolution around the sun.

15. *What is the annual range of beam spreading at the equator?*
    Most direct rays: 1 unit beam = _____ surface units   Date _____
    Least direct rays: 1 unit beam = _____ surface units   Date _____

16. *What is the percentage difference in beam spreading between the highest and lowest amounts at the equator?*

17. *Use the up arrow beneath the right diagram to change the latitude from 23½° N to your latitude or a place north of 23½° N that you'd like to visit. What is the range of beam spreading at this new location?*
    Most direct rays: 1 unit beam = _____ surface units   Date _____
    Least direct rays: 1 unit beam = _____ surface units   Date _____

18. *What is the percentage difference in solar intensity between the highest and lowest amounts at this new latitude?*

19. *Make a general statement about the relationship between latitude and the seasonal range in beam spreading (solar intensity). How does this relationship help to explain why the annual temperature range in the tropics is different than in high latitudes?*

20. Use the top arrows to change the date to the December solstice and record the beam spreading (surface units) at each latitude below. Then change the date to the June solstice and record beam spreading.

	December Solstice	June Solstice
60° N	_____	_____
50° N	_____	_____
40° N	_____	_____
30° N	_____	_____
20° N	_____	_____

21. What do your results reveal about seasonal contrasts in the gradient of solar intensity across the mid-latitudes?

Review Questions

Review your June 21 and December 22 sun-angle calculations for New Orleans and Helsinki. The seasonal sun-angle difference should be the same for both stations—exactly equal to 47°, the difference between the Tropic of Cancer (23½° N) and the Tropic of Capricorn (23½° S). If this is true, then why are seasonal differences in solar intensity (Table 2-1) so much greater at Helsinki?

From the examples you've seen in this lab, describe how seasonal changes in sun angle and daylight hours at a given place will influence its annual temperature range.

GeoClock

Introduction

Why might a region be referred to as the "land of the midnight sun?" Why do places on the equator receive twelve hours of daylight every day of the year? This lab uses GeoClock, a computer program written by Joseph R. Ahlgren, to investigate how earth–sun geometry influences daylight hours.

Running GeoClock

For this exercise you will use the Windows version of GeoClock. When you begin, you will see a graphic similar to the one below. Start the software by clicking on the icon that looks like a miniature of Figure 3-1 (GEOCKWIN.EXE).

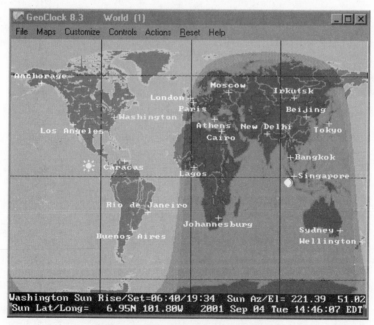

Figure 3-1. GeoClock screen.

Take a minute or two to familiarize yourself with the map and the text at the bottom. The illuminated and dark portions of the earth correspond to the internal clock and calendar on your computer. For a given location, at the specified day and time, the text provides the sun's azimuth and elevation (solar angle), and the latitude and longitude of the sun's direct rays. It also shows the time of sunrise and sunset for a specific location. Note that time is elapsing at a normal rate.

GeoClock Commands

GeoClock commands can be accessed using menus at the top of the screen or by using the keyboard. You will find most of the commands required for this lab under the Maps and Controls menus. Here are some keyboard shortcuts:

Time Control

    T   Allows you to set the date, time, and speed at which time elapses.

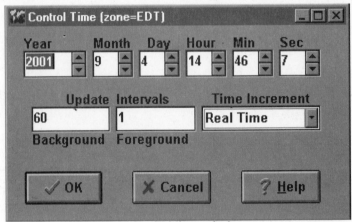

Figure 3-2. Control time window.

Map Selection

    M   Select a map from the list.

    K   Return to previous map display.

Special Commands

    C   Access a command window. A full list of commands is available under the HELP menu.

    V   Access the VCR controls window, to run, stop, and reverse the sun's motion. Clicking the **T** button will also bring up the Control Time window.

    Esc Exit the program.

Figure 3-3. VCR window.

28

If you haven't already done so, start GeoClock.

1. *What is the city shown at the bottom of the screen? The date and the time?*

   Washington   September 24, 2001
                                 1:41 pm

2. *What are the latitude (solar declination) and longitude of the direct solar rays?*   0.69 S   87.53 W

3. *The "Sun Az/El" refers to the position of the sun. Azimuth tells you its direction (90° is east, 180° is south, and 270° is west). Elevation is the solar angle above the horizon. What is the current solar azimuth and elevation for the city shown?*

   Sun Az = 196.41
          El = 49.21

4. *When does the sun rise and set at the city listed? How many hours of daylight does it have?*

   Rise - 6:58 am
   Set - 7:02 pm
   Daylight 12 hrs 4 mins

5. *Solar noon is the midpoint in time between sunrise and sunset. At what time does it occur at this city on this date?*

   1 pm

6. *How would you characterize the shape of the boundary between light and dark portions of the earth? Is a greater proportion of one hemisphere illuminated than of the other?*

   The east is the darker portion of the earth, the shape is a rounded rectangle. It seems that the northern hemisphere is more illuminated

29

To examine daylight hours at different latitudes, it is helpful to draw reference latitude and longitude lines on the globe. The commands below will allow you to draw lines of longitude at 15° intervals that divide the globe (360°) into 24 equal parts. While these lines of longitude do not divide the world into actual time zones, they can be used to estimate daylight hours for various parts of the world since each 15° of longitude represents one hour.

Relevant Commands:

- Access the command window by typing **C** (or choose: Controls | Command entry).
- Enter glinc 15 in the command window and click the OK button.
- Enter **tlinc** 15 in the command window and click the OK button..
- Click the Close button to exit the command window.

7. **Cairo is located at 31° 17' E, Los Angeles at 118° 15' W longitude. What is the difference in longitude between these two cities? _____° Divide this number by 15° to estimate the time difference between the two cities: _____ hours.**

8. **Use the graphic to estimate the daylight hours at three different latitudes on four different dates*:**

		Daylight Hours		
Latitude	Today	June 21	Dec. 22	Mar. 21
_____	_____	_____	_____	_____
_____	_____	_____	_____	_____
_____	_____	_____	_____	_____

Relevant Commands (to change the date):

- Type **V** to open the VCR controls window (or choose: Controls | VCR Controls)
- Click **T** in the VCR controls window (you may want to drag the "Control Time" window away from the main viewing area).
- Enter the appropriate date.
- Click the OK button.

We have examined daylight hours for several dates and places, but have not displayed gradual seasonal changes. In this section, you will view changes in daylight hours at one-day intervals. Read through the questions and Relevant Commands before beginning this section.

Relevant Commands:

In the VCR controls box:

- Click the fast forward button.
- Choose **1d**. (Note that each time the screen is updated, the date changes by one day. Since you are examining the same time each day, the sun remains at the same longitude, but changes latitude each day.)
- Press the pause button to freeze the screen.

9. **Observe the pattern of illumination and the date at the bottom of the screen as the seasonal cycle progresses. When do the number of daylight hours increase in the Northern Hemisphere? When do they increase in the Southern Hemisphere?**

10. **Watch the changes in the pattern of illumination and darkness through the course of a year. When does this pattern, and hence daylight hours, change most rapidly?**

Diurnal Changes

This section shows changes in the illumination of the earth during the course of a day as the earth rotates on its own axis. You will first change the time of year and then adjust the time rate to half–hour intervals. Observe the patterns of daylight change during the course of the day and, more gradually, during the course of the year.

Relevant Commands:

- Choose the Space—N America map (Type **M** or choose: Maps | Map List).
- Click the fast forward button in the VCR Controls box.
- Choose **1h** in the VCR Controls box.

A View From The Poles

In this section you will examine day light hours during each of the solstices and from the perspective of both poles. Read through the questions and the Relevant Commands before continuing.

Relevant Commands:

- Enter **M** to select South Pole from the map list.
- Enter **M** to select the North Pole map.
- Enter **K** to switch between the last two maps.
- Enter **T** to change the dates accordingly.

11. *What can you say about daylight hours in the South Polar Regions around the December solstice? What can you say about daylight hours in the North Polar Regions around the December solstice?*

12. *How are daylight hours in polar regions different around the June solstice?*

Examine seasonal changes at the Poles using the following commands:

Relevant Commands:

- Change the animation time to **1d** in the VCR box.
- Press **K** to toggle between the North Pole and South Pole maps.
- Pause and start the animation as needed.

13. *Toggle between the North and South Pole maps and observe how daylight hours change seasonally at the poles. Freeze the screen when the circle of illumination cuts through the North or the South Pole. When does this occur?*

14. *Freeze the screen when the South Pole region has its longest period of daylight. On which day does this occur?*

The Direction of Sunrise and Sunset

As a final exercise you will examine the patterns of sunrise, sunset, and daylight hours across the United States on the two solstices and the March equinox. Again, you should read through the questions and Relevant Commands first.

Relevant Commands:

- Enter **M** to change the map to United States (48).
- Adjust the dates in the Time Controls window (accessed by typing **T**)
- Choose a time increment of **1h.**

15. *For a given line of longitude does the sun rise and set first in the northern or southern United States? (Circle the correct answer.)*

Date	Rises First	Sets First
Jun. 21	Northern U.S./Southern U.S.	Northern U.S./Southern U.S.
Dec. 22	Northern U.S./Southern U.S.	Northern U.S./Southern U.S.
Mar. 21	Northern U.S./Southern U.S.	Northern U.S./Southern U.S.

16. *The sun does not always rise directly in the east and set directly in the west. Record the azimuth at sunrise and sunset at some location in the United States for the solstices and March equinox. This indicates the part of the sky in which the sun rises and sets on these dates.*

Date	Azimuth at Sunrise	Azimuth at Sunset
June 21	_____	_____
December 22	_____	_____
March 21	_____	_____

Review Questions

So, why might a region be referred to as the "land of the midnight sun"? What is limiting about this description?

Why does the equator always have twelve hours of daylight?

THE SURFACE ENERGY BUDGET

Introduction

This lab introduces you to radiation laws and the fluxes of radiation absorbed, reflected, and emitted by the earth's surface. Using a series of case studies, you will examine seasonal, diurnal, and meteorological influences on solar radiation receipt.

Radiation Laws

All bodies radiate energy. The rate and wavelength of radiation emission depend on the temperature of the radiating body. The *Stefan-Boltzman Law* states that the total energy emitted by a body is proportional to its temperature. Mathematically, we express this law as:

$$E = \sigma T^4$$

where

E = radiation emitted (Wm^{-2})

σ = the Stefan-Boltzman constant ($5.67 \times 10^{-8}\ Wm^{-2}K^{-4}$)

T = temperature (K)

Consider the earth, which has a blackbody temperature of 255 K. How much radiation is emitted by its surface?

$$E = (5.67 \bullet 10^{-8}\ Wm^{-2}K^{-4}) \bullet (255\ K^4)$$

$$E = (5.67 \bullet 10^{-8}\ Wm^{-2}\cancel{K^{-4}}) \bullet (4.228 \bullet 10^9\cancel{K^4})$$

$$E = 240\ Wm^{-2}$$

Note that the earth's average surface temperature is 288 K (15° C). This is 33 K warmer than the emission temperature because the atmosphere effectively absorbs terrestrial radiation, warms, and therefore radiates increasing amounts of energy, some of which is directed towards the surface.

1. **The sun has an average surface temperature of 6000 K. How much radiation is emitted from this surface?**

2. **a. How many times warmer is the sun than the earth?**

$$\left(\frac{solar\ temperature}{earth\ temperature}\right) =$$

   b. What is this number raised to the fourth power.

   c. Does your result approximately equal the ratio of solar/earth emitted radiation?

   Wien's Law states that the maximum wavelength emitted by a body is inversely proportional to temperature. Specifically:

$$\lambda_{max} = \frac{C}{T}$$

   where λ_{max} = the wavelength of maximum emission (micrometers, μ)
             C = Wien's constant (2878 μK)
             T = temperature (K)

3. **Calculate the wavelength of maximum emission for the earth and sun using the emission temperatures above.**

Radiation Fluxes

From the perspective of the surface, the radiation budget is made up of the four fluxes shown in Figure 4-1. The magnitude of each component depends on location, time of year, and state of the atmosphere.

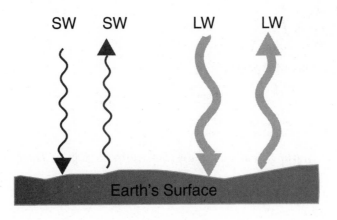

Figure 4-1

SW $\downarrow$ Incoming shortwave radiation measured at the earth's surface.

SW $\uparrow$ Shortwave radiation reflected by the earth's surface.

LW $\downarrow$ Incoming long-wave radiation emitted by the atmosphere.

LW $\uparrow$ Outgoing long-wave radiation emitted by the earth's surface.

Figures 4-2 and 4-3 on the following page show shortwave and long-wave radiation measured at the earth's surface during two days at a site near Barnwell, SC (33½° N latitude). Notice the differences in the four fluxes between July 24 and December 8.

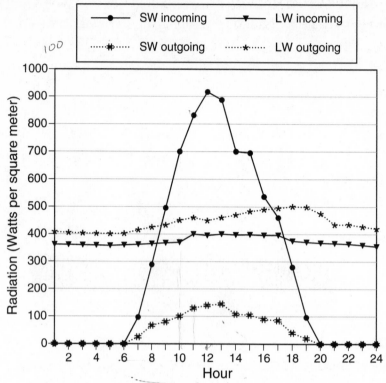

Figure 4-2. Hourly Radiation Fluxes—Barnwell, SC July 24.

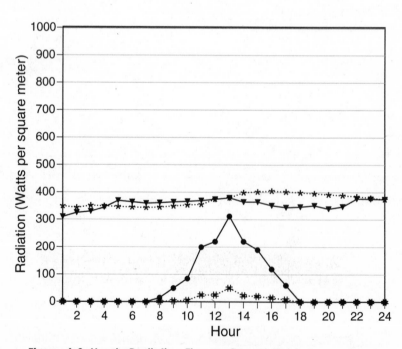

Figure 4-3. Hourly Radiation Fluxes—Barnwell, SC, December 8.

4. Pick _either_ Figure 4-2 _or_ 4-3 to fill in the blanks in the diagram below. Calculate the radiation budget for one daytime hour and one nighttime hour on the day you choose.

Net 700 − 100 = 600

350 − 450 = −120

600 − 120 = 480 w/m²

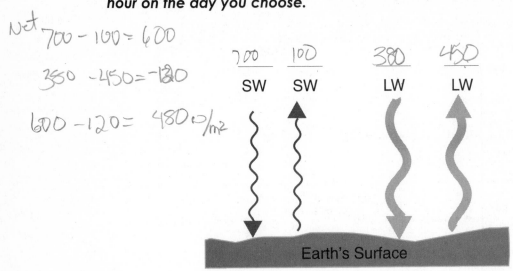

700   100   380   450
SW   SW   LW   LW

Earth's Surface

Daytime: Date _____   Time _____

What is the net surface radiation balance at the time chosen? __480__ Wm⁻²

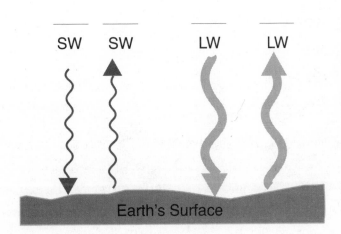

SW   SW   LW   LW

Earth's Surface

Nighttime: Date _____   Time _____

What is the net surface radiation balance at the time chosen? _____ Wm⁻²

5. On July 24 and December 8 the solar noon solar radiation values at the top of the atmosphere above Barnwell are 1317 and 757 watts per meter squared respectively. Why do you think the surface values shown in Figures 4-2 and 4-3 are lower?

6. What is the surface albedo at 1:00 PM on July 24? _____ %

$$(Albedo = \frac{SW\uparrow}{SW\downarrow} \cdot 100\%)$$

7. During which hours of the day is the earth's emitted radiation highest? Why?

8. Why are values of long-wave radiation emitted from the earth higher on July 24 than on December 8?

Radiation and Cloudiness

Figure 4-4 shows the surface fluxes of radiation at the Barnwell site for the four-day period November 9–12, 1983. Changes in each of the fluxes can be explained by considering the radiation laws, the normal diurnal cycle, and changes in atmospheric composition and cloudiness.

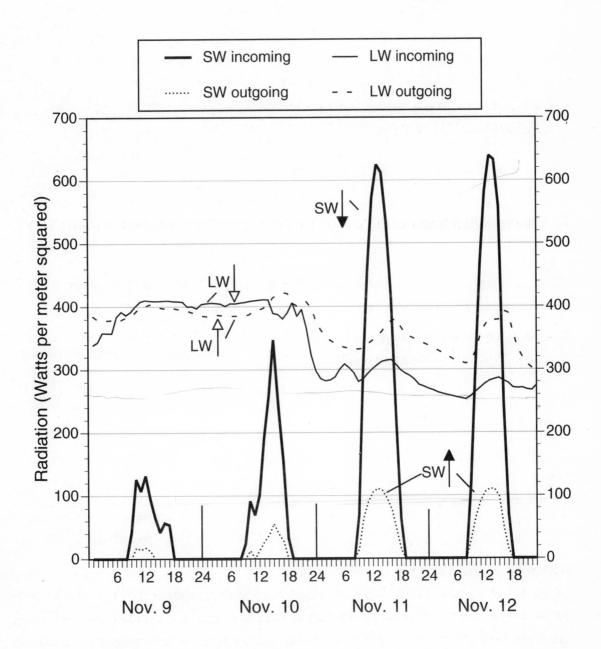

Figure 4-4. Hourly Radiation Fluxes—Barnwell, SC, November 9-12.

9. Based on the amounts of incoming shortwave radiation, which days between November 9 and November 12 were clear and which were cloudy?

10. a. Based on the long-wave radiation fluxes, estimate the day and time when surface temperature was highest.

    b. On which day and at what time was it lowest?

    c. How did you make your estimates?

11. a. On which day and during what hours did the incoming long-wave radiation from the atmosphere change most dramatically?

    b. How might this sharp change be related to changes in cloudiness during the period?

Interactive Computer Exercise: Shortwave Radiation

This applet allows you to examine how cloud cover, latitude, season, and time of day affect incoming shortwave radiation. The questions below focus on the relative importance of each of these factors on solar radiation receipt.

12. Record the amount of incoming solar radiation at an initial setting without cloud cover.

Date:	June 22	Time:	12:00
Latitude:	30° N	Albedo:	15%
Clouds:	High: 0	Medium: 0	Low: 0

    Incoming SW Radiation _____

13. Compare this value with the amount of solar radiation at the top of the atmosphere (Table 2-1). What do you think accounts for the reduction in solar radiation striking the earth's surface?

14. Add some combination of clouds to reduce incoming solar radiation by approximately half. Record your results.

    Date: _____June 22_____   Time: _____12:00_____
    Latitude: ___30° N___   Albedo: _____15%_____
    Clouds: High: _____ Medium: _____ Low: _____
    Incoming SW Radiation _____

15. Are the effects of cloud cover the same regardless of cloud height?

16. Eliminate all clouds, then adjust the date to see if you can reduce the incoming solar radiation by half. If you can, record the date; if not, make a note of this. Would you get the same result for a high latitude?

    Date: _____   Time: _____12:00_____
    Latitude: ___30° N___   Albedo: _____15%_____
    Clouds: High: _0_ Medium: _0_ Low: _0_
    Incoming SW Radiation _____

17. By how much would you have to change the time of day to reduce incoming solar radiation by half?

    Date: _____June 22_____   Time: _____
    Latitude: ___30° N___   Albedo: _____15%_____
    Clouds: High: _0_ Medium: _0_ Low: _0_
    Incoming SW Radiation _____

18. *How does albedo influence the net radiation (incoming shortwave radiation minus reflected shortwave radiation)at the earth's surface?*

Interactive Computer Exercise: Long-wave Radiation

This applet will let you explore the relationship between the earth's surface temperature and its emission of long-wave radiation. It also includes a simplified treatment of atmospheric long-wave emission back towards the earth's surface. Start the applet and answer the questions below.

19. *How does long-wave radiation emission from the earth's surface change with increasing temperature?*

20. *Note that the difference in surface emission between -10° C and -5° C is not the same as the difference in surface emission between 35° C and 40° C. Use one of the radiation laws to explain why this is the case.*

21. *How does simulated incoming long-wave radiation change with increasing cloudiness? How does the amount of incoming long-wave radiation depend on cloud height?*

The applet provides a simplified treatment of long-wave emission from the atmosphere which is dependent only on cloud amount. In reality, atmospheric temperature will determine how much radiation is radiated back towards the earth's surface.

22. *What factors besides clouds could alter atmospheric temperature and, therefore, atmospheric emission of radiation?*

23. *Cloudy nights do not cool as fast as clear nights. How does this applet illustrate the processes causing this? How does the applet limit consideration of this phenomenon (i.e., is there a variable in the applet that should change in response to a radiation flux, but does not)?*

Review Questions

Which varies more seasonally, incoming shortwave (solar) radiation or outgoing long-wave (terrestrial) radiation? Which varies more diurnally? Explain your answers.

What information would you want to know in order to predict tomorrow's high and low temperatures? How would this information help you?

Simulating the Earth's Energy Budget

Introduction

In Lab 4 we examined radiative fluxes at a local scale. Here we will consider the global energy budget. This lab uses a one-dimensional energy balance model, developed by Professor James E. Burt, to simulate radiation fluxes in the earth-atmosphere system. The model predicts surface temperature as a function of the variables shown in the table below.

Running the Model

Your instructor will provide basic instructions to run the model. The model runs under DOS, so you must open a DOS window and start the model by typing **ebmodel**. This command will lead you to the main menu showing the model's basic parameters and settings:

Type of Run: GLOBAL RUN		
Parameter Name	Current Setting	Change from Normal
Solar Constant	1367.0 W/m^2	0.0%
Atmospheric Transmissivity	0.890	0.0%
Carbon Dioxide Concentration	350.0 ppm	0.0%
Global Surface Albedo	0.121	0.0%
Albedo Feedback	0.000 / °C	0.000 / °C
Eddy Diffusion Factor	1.0	0.0%
Ocean Mixed Layer Depth	75.0 meters	0.0%
Year	1990.0 AD	0.0 years
Eccentricity	0.017	0.0%
Day of Perihelion	2.8	0.0 Days
Obliquity	23.45°	0.00°
Run e**X**it **G**raphs **T**ables re**S**et Info		

Throughout this lab you will be adjusting these settings to simulate the climatic response to specific changes. Move through the model's menu with the arrow keys. As you do, notice that the message in the bottom panel changes to correspond to each of the parameters chosen. You may change the value of any parameter at the ">" prompt.

Constructing Graphs and Tables

Many of the following exercises will require examination of graphs and tables. The **Graphs** and **Tables** commands at the bottom of the main menu will bring you to graph and table submenus. The same commands will work in both of these submenus.

GRAPH SUBMENU

Variable to Plot	Results from Seasonal Run			
		Monthly Plot for:		Zonal Plot for:
A Temperature				
B Incident Solar Radiation	A 85S	J 85N	1 Jan	7 Jul
C Absorbed Solar Radiation	B 75S	K 75N	2 Feb	8 Aug
D long-wave Radiation	C 65S	L 65 N	3 Mar	9 Sep
E Northward Heat Transport	D 55S	M 55N	4 Apr	10 Oct
F Albedo	E 45S	N 45N	5 May	11 Nov
	F 35S	O 35N	6 Jun	12 Dec
X Exit Plot Menu	G 25S	P 25N		
	H 15S	Q 15 N		
	I 5S	R 5N		
On all plots:				
	S Plot Global Mean		13 Plot Annual	
solid lines—results from	by Month		Mean by Zone	
most recent model run				
(if available)	Results from Annual Run			
dotted lines—results under	14 Plot Annual Mean by Zone			
normal conditions				
Enter variable to plot (A-X) > **C**	Enter zone (A-S) or month (1-14) > **L**			

Monthly/Zonal Tables	Results from Seasonal Run		
	Monthly Plot for:		Zonal Plot for:
A Temperature	A 85S	J 85N	1 Jan 7 Jul
B Incident Solar Radiation	B 75S	K 75N	2 Feb 8 Aug
C Absorbed Solar Radiation	C 65S	L 65 N	3 Mar 9 Sep
D long-wave Radiation	D 55S	M 55N	4 Apr 10 Oct
E Northward Heat Transport	E 45S	N 45N	5 May 11 Nov
F Albedo	F 35S	O 35N	6 Jun 12 Dec
	G 25S	P 25N	
X Exit Plot Menu	H 15S	Q 15 N	
	I 5S	R 5N	
Other Tables	S Plot Global Mean	13 Plot Annual	
	by Month	Mean by Zone	
G Global Summaries			
P Parameter Values			
N New Printer Port	Results from Annual Run		
X Exit Print Menu			
	14 Plot Annual Mean by Zone		
Printing to lpt1:			
Enter variable to plot (A-X) > **C**	Enter zone (A-S) or month (1-14) > **L**		

Two selections must be made to plot a graph or display a table:

- Select the letter corresponding to the variable you'd like to examine (e.g., **A** for Temperature, **B** Incident Solar Radiation, **C** Absorbed Solar Radiation).
- Select a yearly table for a particular latitude (e.g., **A** for 85° S, **B** for 75° S, **C** for 65° S) *or* select a zonal table for a given month (e.g., **1** for January, **2** for February, **3** for March).
- To return to the graph or table submenu: x **<enter>**
- To exit the graph or table submenu: x **<enter>**

Most of the following exercises will focus on seasonal changes in the energy budget. Use the up arrow key (↑) to move to **Type of Run** at the top of the main menu. Change the type of run to **Seasonal Run** by using the right arrow key (→).

1. *Type "R" to run the model. What is the global annual average temperature?*
    _____° C

Absorbed Solar Radiation

The amount of solar radiation absorbed at the surface depends on sun angle and surface albedo, both of which change seasonally.

2. *Use the software to plot graphs of absorbed solar radiation for each latitude shown in the table below. Record the June and December absorbed solar radiation and compute the seasonal range. (You can find precise values for each latitude and month using the Tables Menu, but it is also useful to see the seasonal patterns graphically.)*

Latitude	Absorbed Shortwave Radiation (Wm²)		
	June	December	Seasonal Range
65° N	293	0	293
35° N			
5° N			
35° S			
65° S			

3. *At which latitudes do you find the greatest seasonal range in absorbed solar radiation? Which two factors account for the large seasonal range there?*

4. *Describe and explain the seasonal pattern of absorbed shortwave radiation at 5° N latitude. During which months is absorbed shortwave radiation highest? During which months is it lowest? What accounts for these patterns and*

for the fact that there are two peaks of absorbed solar radiation at 5˚ N latitude and only one at the other latitudes considered?

5. *Why is incident shortwave radiation in December at 35˚ S higher than incident shortwave radiation in June at 35˚ N?*

Temperature

6. *Plot graphs of temperature for each latitude shown in the table below. Complete the table by noting the warmest and coolest months and their respective values.*

| | Temperature | | |
Latitude	Warmest Month	Coolest Month	Seasonal Range
65˚ N	August: 15˚ C	March: -13˚ C	28˚ C
35˚ N			
5˚ N			
35˚ S			
65˚ S			

7. *Where is the seasonal temperature range greatest? Why is it greatest at this location?*

8. *Why is the seasonal temperature range greater at 35˚ N than at 35˚ S?*

Terrestrial Long-Wave Radiation

The earth emits radiation skyward. The amount of radiation leaving the earth-atmosphere system varies with the temperature of the earth and the composition of the atmosphere.

9. *Plot graphs of long-wave radiation for each latitude shown in the table below. Complete the table by indicating the months of highest and lowest long-wave radiation and the amounts emitted during these months. Also record the long-wave radiation during June and December.*

	Long-wave Radiation (Watts/meter^2)		Highest	Lowest
Latitude	June	December	Month Amount	Month Amount
65° N	208	181	August: 240	March: 159
35° N				
5° N				
35° S				
65° S				

10. *Compare your results from the temperature table with those of the long-wave radiation table. Explain the relationship between temperature and emitted long-wave radiation with respect to one of the radiation laws.*

Net Radiation

A radiation budget can be computed for any location by measuring the amount of absorbed shortwave radiation and subtracting from it the amount of outgoing long-wave radiation.

11. *Calculate the net radiation during June and December at each latitude.*

	Net Radiation	
Latitude	June	December
65° N	293 - 208 = 85	0 - 181 = -181
35° N		
5° N		
35° S		
65° S		

Northward Heat Transport

The earth's atmosphere and oceans circulate excess energy from tropical regions poleward. Therefore, the net energy budget of a given latitude is more than a function of absorbed shortwave and emitted long-wave radiation; it also depends on receipt or loss of energy from other latitudes. The model allows you to examine energy transport between latitudes by selecting the variable "Northward Heat Transport" from the graph submenu. Positive values indicate northward energy transport; negative values indicate southward heat transport.

12. *Over which Northern Hemisphere latitudes do you find the greatest heat transport during June? Where is the greatest transport during December? What accounts for the seasonal difference?*

Heat Storage

The annual temperature peak at most mid-latitude locations occurs approximately one month after peak absorbed shortwave radiation. Similarly, minimum temperature lags about one month behind minimum absorbed shortwave radiation. The lags occur because absorbed solar energy is not re-emitted immediately, but is stored at the earth's surface and slowly released to the atmosphere. In the model you can alter the storage capacity by changing the ocean mixed layer depth. By default, the model stores energy in the top 75 meters.

- Adjust the "Ocean Mixed Layer Depth" parameter to **25** meters. (The model will now store energy only in the top 25 meters of the ocean.)
- Run the model and graph the temperature for **35° N** latitude.

13. *How does this seasonal march of temperatures differ from the standard model run (shown in red) which stores energy in the top 75 meters? Explain why the differences occur.*

- Set the "Ocean Mixed Layer Depth" back to **75** meters.
- Run the model.

Albedo

Albedo, or reflectivity, at the earth's surface is a function of sun angle and of the color and texture of the surface. Albedo values range from 0% (no reflection) to 100% (complete reflection).

14. *Plot albedo for 65° N latitude. What is the seasonal range of albedo at this latitude? What accounts for the seasonal differences?*

15. Note the zonal plot of January albedo shown in Figure 5-1. Use the same figure to plot albedo for July. Compare it to the albedo plot for January. How do the patterns differ? What accounts for these differences?

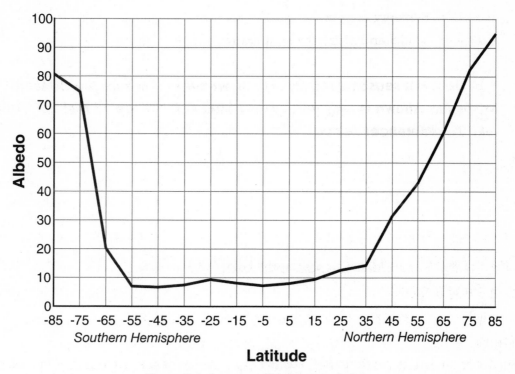

Figure 5-1. January albedo—zonal plot.

16. Return to the main menu; adjust the global albedo to 0.200. Run the model and record the global annual average temperature. _____°C.

17. How and why does a change in albedo affect the earth's temperature?

55

Review Question

Summarize the earth's energy budget by explaining how a given latitude can gain energy and how it can lose energy.

ATMOSPHERIC MOISTURE

Materials Needed
- calculator
- ruler

Introduction
What is humidity and how is it important to atmospheric processes? This lab shows how we measure atmospheric moisture and uses two experiments to illustrate its role in energy transfer.

Latent Heat
Water is a unique substance because it occurs in three phases (solid, liquid, and vapor) at temperatures and pressures that commonly occur on earth. The phase change of water is what makes the weather interesting, variable, and often unpredictable. When it changes phase, water releases or consumes energy—a process called *latent heat transfer*. The following two experiments illustrate the concept of latent heat. If you do not have the necessary equipment to conduct these experiments, please read through them anyway, carefully review the section on latent heat in your textbook, and answer the general questions (10–13).

Optional Exercise: Energy Transfer Experiments

EXPERIMENT I
Procedures

A. Fill a beaker with 300 ml of water and heat it to a boil. While the water is heating, complete step B.

B. Fill another beaker about ¾ full of ice and the remainder with water. Stir for 1 minute and record the temperature.

Ice-water Temperature = _____ ° C

C. Put 50 ml of the heated water and a thermometer in a beaker. When the tempera-
    ture of the water has dropped to 80° C, add to it 50 ml of the water from the
    ice-water mixture in step B. Stir and record the final temperature.

    Final Temperature = _____ ˚ C

Calculations

Note: A *calorie* is the energy required to raise the temperature of one gram of water
one degree centigrade. One ml of water weighs 1 gram since the density of water is
$1 \text{ g}/\text{cm}^3$, and $1 \text{ ml} = 1 \text{ cm}^3$.

1. **How many calories of heat were lost by the 50 ml of hot water as it cooled
    from 80˚ C to the final temperature?**

    _____ calories

2. **How many calories were gained by the cold water?**

3. **Was the energy lost by the hot water nearly equal to the amount gained by
    the cold water? Should it be?**

EXPERIMENT II

Procedures *(Steps A & B should be done concurrently)*

A. Chill an empty beaker. Weigh it and add 25 grams of ice. Then pour water from the
    ice-water beaker (from Experiment I) into the first beaker until it contains a total of
    50 grams of ice and water. Record the temperature of the contents.

    Ice-water Temperature = _____ ˚ C

B. Pour 50 ml of boiling water into another beaker and allow it to cool to 80° C.

C. Pour the 80° C water into the ice-water mixture and stir. After all the ice has melted
    record the temperature.

    Temperature = _____ ˚ C

CLOUD DROPLETS AND RAINDROPS

Materials Needed
- calculator

Introduction

How do raindrops and snowflakes form? This lab examines the forces acting on cloud droplets and ice crystals and the processes that cause these droplets and crystals to grow. We will consider how such growth influences the probability that clouds will produce precipitation.

Cloud Droplet Growth

Clouds form as water vapor either condenses on condensation nuclei or is deposited on freezing nuclei. Typically, cloud droplets and ice crystals are very small (50 μm = 0.05 mm). For clouds to produce precipitation, cloud droplets must grow large enough and fast enough to fall to the ground without evaporating. Droplets reach this size through two processes: the *Bergeron process* and the *collision-coalescence process*. The Bergeron process of ice-crystal growth depends on the coexistence of water vapor, supercooled liquid water droplets, and ice. To understand the importance of this coexistence, consider again the Clausius-Clayperon (saturation vapor pressure) curve in Figure 8-1. Remember that the solid curve depicts the equilibrium point between liquid water and atmospheric water vapor over a range of temperatures. An air sample below the curve is considered unsaturated, indicating that evaporation can readily occur, and an air sample above the curve is considered *supersaturated*, suggesting that condensation will likely occur.

Clouds with temperatures between -10° and -20° C often contain water in all three forms—water vapor, supercooled liquid water droplets, and ice crystals. The tendency for water to move from one phase to another depends on the equilibrium between phases. Since water molecules are more tightly bonded in solid than in liquid form, more energy is required to change the phase of water from solid to vapor (a process called *sublimation*) than from liquid to vapor (evaporation). Therefore, the saturation vapor pres-

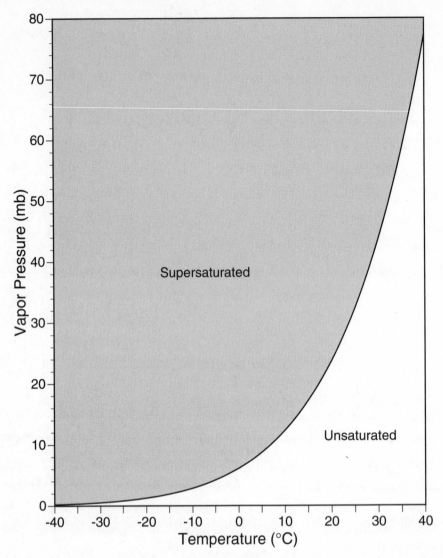

Figure 8-1. Saturation vapor pressure.

sure over liquid water is greater than the saturation vapor pressure over ice and it is possible for an air sample to be unsaturated with respect to liquid water and supersaturated with respect to ice (Figure 8-2).

1. Consider the air sample denoted by a "*" in Figure 8-2. Its temperature is -15° C and its vapor pressure is 1.72 mb. What is its relative humidity with respect to water if, at -15° C, the saturation vapor pressure over liquid water is 1.91 mb? (Your answer should be less than 100%, indicating an unsaturated sample.)

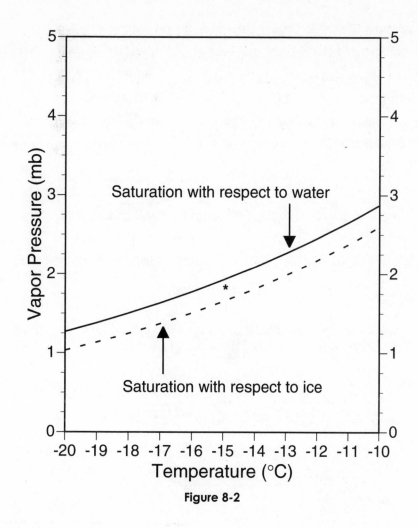

Figure 8-2

2. **What will likely happen to liquid water when the surrounding environment is unsaturated?**

3. **What is its relative humidity with respect to ice if, at -15° C, the saturation vapor pressure over ice is only 1.65 mb?**

4. **What happens to water vapor if, with respect to the surrounding ice crystals, the environment is supersaturated?**

Forces Acting on Cloud Droplets and Raindrops

Two opposing forces act on droplets—*gravity* and *friction*. Gravitational force pulls a water droplet to the earth's surface and is equal to the mass of the droplet times gravitational acceleration (Gravity = Mass $\times$ 9.8 m $\bullet$ sec^{-2}). Since one cubic centimeter (1 cm^3) equals 1 gram, we can substitute volume for mass in the above equation. Therefore, the gravitational force of a droplet increases with its volume. The volume of a sphere is calculated as follows:

$$V = \frac{4}{3} \pi r^3$$

where r is the radius of the droplet, equal to ½ its diameter.

5. *What is the volume of an ordinary cloud droplet (25 µm = 0.025 mm radius)?*

6. *What is the volume of a small raindrop (500 µm = 0.5 mm radius)?*

7. *A large raindrop (2500 µm = 2.5 mm radius) is 100 times the radius of a typical cloud droplet (25 µm). How many times greater is the volume of a large raindrop?*

As a droplet falls it encounters air resistance, or frictional force. The magnitude of this force depends on the size of the drop's "bottom"—i.e., the surface area resisting the fall. Figure 8-3 illustrates the "bottom" of a cloud droplet with 25 µm spherical radius. Assuming the drop is spherical, the surface area experiencing friction is that of a circle, calculated as: Area = πr^2.

8. *What is the area of the "bottom" of a falling droplet of 1-mm diameter?*

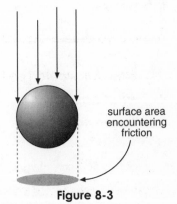

surface area encountering friction

Figure 8-3

9. How many times greater is the "bottom" area of a falling droplet of 5-mm diameter?

10. Which of the two forces, gravity or friction, increases more rapidly as droplet size increases?

Terminal Velocity

Frictional drag increases as a droplet accelerates, since a faster droplet encounters more air molecules. Eventually frictional and gravitational forces balance and the droplet no longer accelerates, but falls at constant speed. This speed is referred to as the droplet's *terminal velocity*. Your calculations above show that as droplet size increases, gravitational force is affected more than frictional force. Therefore, smaller droplets will have a lower terminal velocity than larger droplets (Table 8-1). Terminal velocity is important because the speed of a droplet's fall determines the probability that the droplet will reach the surface before evaporating. Droplets with radii smaller than 75 μm are unlikely to reach the ground because they fall so slowly.

TABLE 8-1

The Relationship Between Droplet Size and Terminal Velocity

Radius (μm)	Rate of Fall (meters per second)	Type of Drop
2500 (2.5 mm)	8.9	Large Raindrop
500	4.0	Small Raindrop
250	2.8	Fine Rain or Large Drizzle
100	1.5	Drizzle
50	0.3	Large Cloud Droplet
25	0.076	Ordinary Cloud Droplet
5	0.003	Small Cloud Droplet
0.5	0.00004	Large Condensation Nucleus

11. How far would the large cloud droplet fall before evaporating?

Since we see clouds because sunlight reflects off individual droplets, the relatively short fall distance of even large cloud droplets contributes to the perception of a sharp cloud boundary.

12. *In theory, how long would it take a large (50 µm radius) cloud droplet to hit the ground if falling from a cloud base at 2000 meters?*

13. *How long would it take a large raindrop (2500 µm radius) to reach the ground if falling from a cloud base of 2000 meters?*

In reality, it is unlikely that a droplet with a 50-µm radius would make it to the surface without evaporating. Table 8-2 lists the maximum fall distance of various droplets before evaporation.

TABLE 8-2

Drop radius (µm)	Maximum fall distance before evaporation (m)
2500	280,000
1000	42,000
500	1,000
200	500
100	150
50	0.1
10	0.033
2	0.00002
1	0.0000033

14. *Approximately how far could drizzle-sized droplets fall before evaporating?*

Figure 8-4 shows how cloud base height influences the probability that precipitation (if occurring) will reach the ground as drizzle or rain. The solid curve shows the probabilities for a low cloud base; the dashed curve shows the probabilities for a relatively high cloud base.

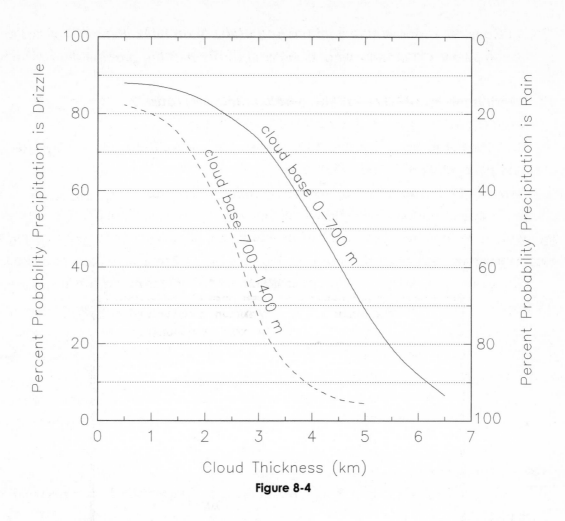

Figure 8-4

15. *If precipitation falls from a cloud 1 km thick with a low base, what is the probability that it will be drizzle? _____%*
What is the probability that it will be rain? _____%

16. *If precipitation falls from a 5.5 km-thick cloud with a low base, what is the probability that it will be drizzle? _____%*
What is the probability that it will be rain? _____%

17. *Why does the probability that the precipitation will be rain and not drizzle increase with a thicker cloud?*

18. a. Consider precipitation from a cloud that is 3 km thick. If the cloud has a base below 700 meters, what is the probability that the precipitation will be drizzle? _____%
What is the probability that the precipitation will be rain? _____%

b. If the cloud base is 700 to 1400 meters high, what is the probability that the precipitation will be drizzle? _____%
What is the probability that the precipitation will be rain? _____%

19. Explain why precipitation from the higher cloud base is less likely to strike the ground as drizzle than precipitation from the lower cloud base.

Review Questions

Why does cloud droplet and ice crystal size influence the probability that clouds will produce precipitation?

How do the height and thickness of clouds influence the probability of precipitation?

WEATHER MAP ANALYSIS

Introduction

With a few exceptions (e.g., clouds), most atmospheric processes are invisible. How then do we "see" the weather in order to forecast its changes? The purpose of this lab is to learn how to construct and interpret weather maps. We will focus on the mid-latitudes where identification of air masses, fronts, and mid-latitude cyclones can help meteorologists forecast changing weather patterns.

Surface Weather Maps

Every six hours atmospheric data are collected at approximately 10,000 surface weather stations around the world. These data are transmitted to one of three World Meteorological Centers in Melbourne, Australia; Moscow, Russia; or Washington, D.C. Weather data are disseminated to national meteorological centers where *synoptic-scale* maps are generated. *Synoptic* literally means coincident in time and a synoptic map is a map of weather conditions for a specific time. By convention, the time printed on many weather maps is Greenwich Mean Time (GMT, also called Universal Coordinated Time), the time at the Prime Meridian. Meteorologists often call this *Zulu* or *z* time. Thus, a map labeled 1200z shows conditions at noon in London, which is 7:00 AM EST in New York.

In the United States an automated weather network collects hourly surface data. Since each station collects data for as many as eighteen weather characteristics, a compact method of symbolization must be used to include all this information on a single weather map. The station model, developed by the World Meteorological Organization, is the standard format for symbolizing weather characteristics. Figure 10-1 illustrates the arrangement of data in the WMO model; Appendix C provides a complete list of symbols used in this lab.

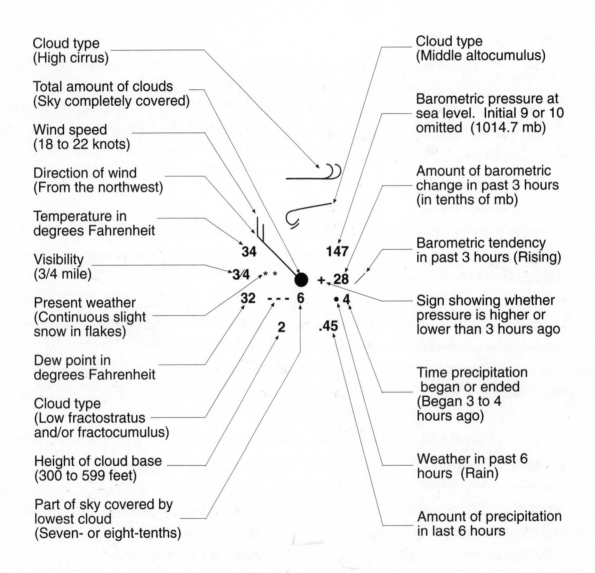

Cloud type
(High cirrus)

Total amount of clouds
(Sky completely covered)

Wind speed
(18 to 22 knots)

Direction of wind
(From the northwest)

Temperature in
degrees Fahrenheit

Visibility
(3/4 mile)

Present weather
(Continuous slight
snow in flakes)

Dew point in
degrees Fahrenheit

Cloud type
(Low fractostratus
and/or fractocumulus)

Height of cloud base
(300 to 599 feet)

Part of sky covered by
lowest cloud
(Seven- or eight-tenths)

Cloud type
(Middle altocumulus)

Barometric pressure at
sea level. Initial 9 or 10
omitted (1014.7 mb)

Amount of barometric
change in past 3 hours
(in tenths of mb)

Barometric tendency
in past 3 hours (Rising)

Sign showing whether
pressure is higher or
lower than 3 hours ago

Time precipitation
began or ended
(Began 3 to 4
hours ago)

Weather in past 6
hours (Rain)

Amount of precipitation
in last 6 hours

34

3⁄4

32

147

+.28

.4

.45

6

2

Figure 10-1. WMO station model.

116

THUNDERSTORMS AND TORNADOES

Introduction

What causes thunderstorms and tornadoes? These dramatic events develop when warm, moist air is forced to rise in an unstable atmosphere. Thunderstorms are often categorized as *air mass thunderstorms* or *severe thunderstorms*. The distinction recognizes that some develop within a warm, moist air mass, while others require strong vertical wind shear. We will focus on the geographic patterns of these storms, their structure, and the atmospheric conditions that produce them.

Spatial Patterns of Thunderstorms

In the United States, thunderstorms most frequently occur when maritime tropical air is forced aloft through orographic lifting, frontal lifting, intense surface heating, or convergence of surface air flow. Figure 12-1 shows the pattern of thunderstorm frequency in the United States.

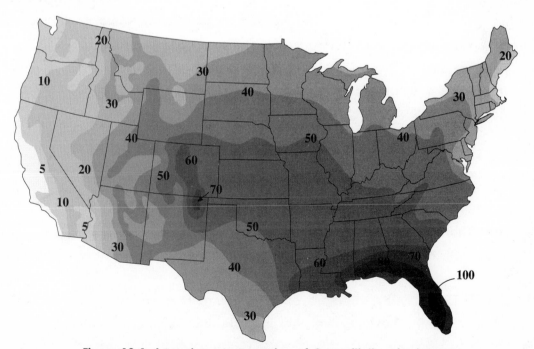

Figure 12-1. Annual average number of days with thunderstorms.

149

1. *What is the major source of maritime tropical air affecting the United States? How does proximity to this source influence the spatial pattern of thunderstorms?*

2. *What processes over the Florida peninsula force maritime tropical air to rise and lead to the high frequency of central Florida thunderstorms?*

3. *What type of local wind might account for the high frequency of thunderstorms in central Colorado and northeastern New Mexico? Explain how such winds could contribute to thunderstorm development.*

Lightning

Lightning is present in all thunderstorms. While it can be seen over great distances, the resulting thunder usually cannot be heard beyond 30 kilometers. Since light travels substantially faster than sound, we see a lightning flash before we hear the accompanying thunder. In fact, we see lightning nearly instantaneously, while sound travels at approximately 330 meters per second (1100 ft/sec; or roughly one-fifth of a mile per second).

4. *If you see a lightning bolt in the distance and 15 seconds elapse before you hear the corresponding thunder, how far away was the lightning bolt?*
   _____ meters   _____ miles

Air Mass Thunderstorms

Air mass thunderstorms arise when air becomes unstable within an already warm, moist air mass. This most commonly occurs during the warmer months, when intense surface heating creates instability within maritime tropical (mT) air masses.

We can describe the development of these thunderstorms in three stages: cumulus, mature, and dissipating. In the cumulus stage, strong updrafts dominate the circulation

150

within a cumulonimbus cloud, causing the cloud to grow vertically. During the mature stage, precipitation begins and downdrafts develop side by side with updrafts. Cooler, drier air from outside the thunderstorm is drawn into the cloud in a process called *entrainment*. Eventually downdrafts dominate the circulation and the storm begins to dissipate. Figure 12-2 shows features of these three stages.

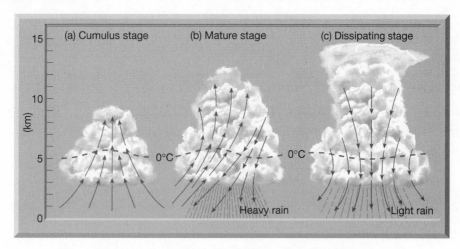

Figure 12-2. Stages in thunderstorm development.

5. **Recall that updrafts within the thunderstorm result in moist adiabatic cooling. How is energy released to the cloud in this process and how might this foster more upward motion within the cloud?**

6. **The air drawn into a thunderstorm from the outside is usually cooler than the air within a cumulonimbus cloud. Explain what effect this should have on vertical motion.**

7. *Entrained air is drier than the air within a cloud. How could this lead to further cooling?*

8. *How could updrafts and downdrafts contribute to cloud droplet or ice crystal growth?*

Severe Thunderstorms

Severe thunderstorms produce frequent lightning and damaging winds or hail. Like air mass thunderstorms, they need warm, moist, conditionally unstable air at the surface. In contrast to air mass thunderstorms, however, severe thunderstorms require upper-air divergence and strong vertical wind shear. These conditions often exist in association with features of mid-latitude cyclones and are very common in the central United States during the spring and early summer. At this time, the contrast between maritime tropical air from the Gulf of Mexico and continental polar air from the north is largest, and consequently the jet stream is strongest. Along a cold front boundary, warm moist air can be forced upward by the intruding cold air. Its ascent is fostered by the jet stream which produces upper-air divergence and vertical wind shear.

Occasionally, a line of thunderstorms develops in the maritime tropical air mass ahead of a surface cold front. Such *squall lines* result from upper-air divergence associated with a strong jet stream and gravity waves triggered by the advancing cold front. The line of thunderstorms is often maintained through self propagation. Figure 12-3 shows a cross-section of a squall-line thunderstorm. It illustrates how the downdrafts of a mature thunderstorm may trigger the development of a new convective cell.

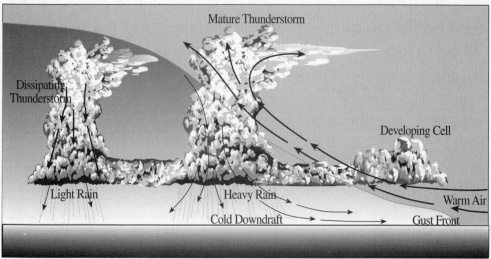

Figure 12-3. Cross-section of a squall-line thunderstorm.

9. *Explain how downdrafts can trigger the development of new thunderstorm cells.*

10. *How do air mass conditions ahead of the squall-line support the development of new cells?*

Severe thunderstorms in the United States also form along *dry lines*. These features develop most frequently in the Southern Plains where continental tropical and maritime tropical air masses often meet. Because continental tropical air develops in the Southwest, where elevations are relatively high, it flows over maritime tropical air when both air masses move into the south-central U.S. The continental tropical air can actually form a capping upper-air inversion that suppresses all but the strongest updrafts. This phenomenon is discussed further with respect to tornado development.

11. *Label the cP, cT, and mT air masses in the map below (Figure 12-4). Use the conventional symbol to draw a cold front separating the cP and mT air masses, and a dashed line to show a dry line separating cT and mT air masses.*

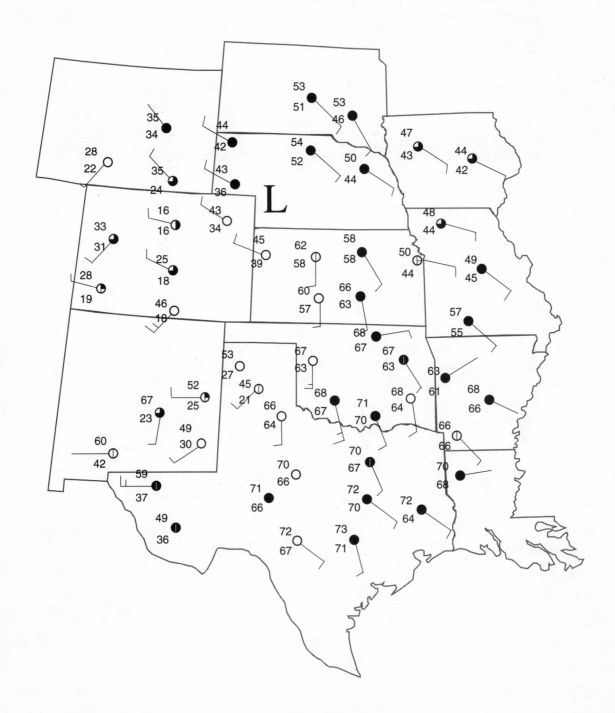

Figure 12-4

Tornadoes

Tornadoes occur in conjunction with severe thunderstorms and, therefore, are associated with the same conditions. The next series of exercises uses a tornado outbreak in Illinois, Indiana, Ohio, and Kentucky on June 2–3, 1990 to illustrate how certain surface and upper-air conditions can lead to severe weather. Figure 12-5 shows the paths of seven tornadoes that occurred from 5:45–11:10 PM CDT on June 2nd. These tornadoes were categorized as F4 on the Fujita Intensity Scale, indicating wind speeds of 333–419 kilometers per hour (207–260 mph). They were the most devastating of fifty-five documented tornadoes occurring in the region that day.

Figure 12-5. F4 tornadoes on June 2, 1990.

Figures 12-6 through 12-9 show the synoptic conditions for June 2nd and 3rd at 7:00 AM CDT (approximately 12 hours before and after the tornado outbreak).

12. **List the <u>June 2</u> surface weather conditions at two stations in the region of the tornadoes.**

Station: _____   Station: _____

Temperature: _____   Temperature: _____

Dew Point: _____   Dew Point: _____

Wind Direction: _____   Wind Direction: _____

13. **How did the surface conditions (temperature, dew point, wind direction) contribute to the day's severe weather?**

14. **Compare temperature and dew point at stations east and west of the cold front on June 2. How sharp was the contrast in air masses on this date?**

15. **How did the change in 500-mb winds between 7:00 AM CDT June 2 and 7:00 AM CDT June 3 contribute to the severe weather?**

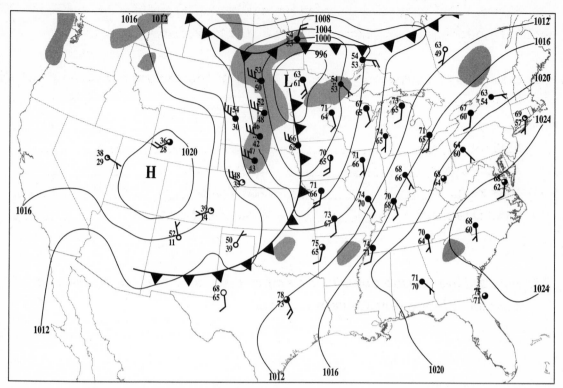

Figure 12-6. Surface conditions, Saturday, June 2, 1990, 7:00 AM CDT.

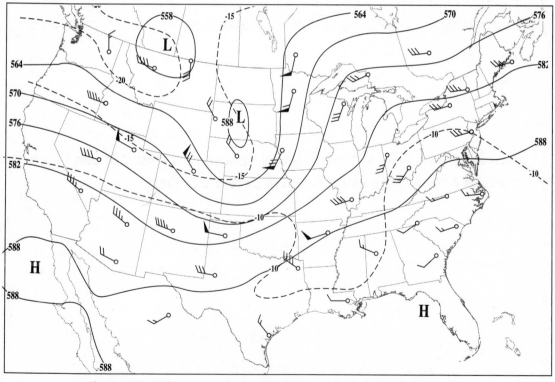

Figure 12-7. 500-mb contours, Saturday, June 2, 1990, 7:00 AM CDT.

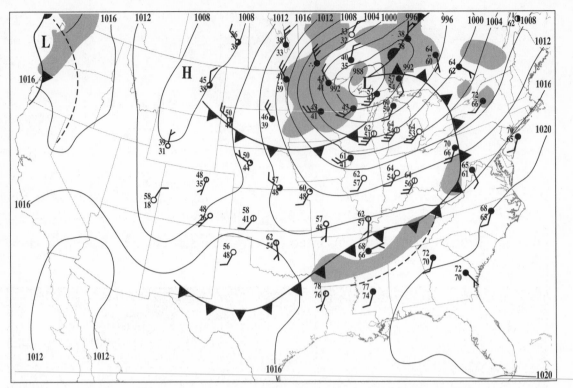

Figure 12-8. Surface conditions, Sunday, June 3, 1990, 7:00 AM CDT.

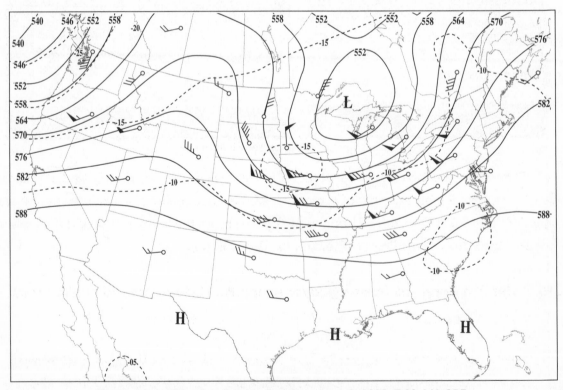

Figure 12-9. 500-mb contours, Sunday, June 3, 1990, 7:00 AM CDT.

Upper-air data at individual sites provide information that is useful for severe thunderstorm prediction. Among other clues, meteorologists examine temperature and moisture characteristics at various heights, determine the stability of the atmosphere, and search for upper-air inversions and wind shear.

An *upper-air inversion* is defined as a layer above the earth's surface in which temperature increases with height. Upper-air inversions foster severe weather by temporarily suppressing vertical mixing between warm, moist air near the surface and cool, dry air aloft. This "cap" prevents vertical cloud development. It can form when a layer of air subsides and warms adiabatically, or when continental tropical air is advected aloft. However, when triggered by a cold front or squall line, warm air can be forced into the stable inversion layer, resulting in intense convection and tall cumulonimbus clouds (Figure 12-10).

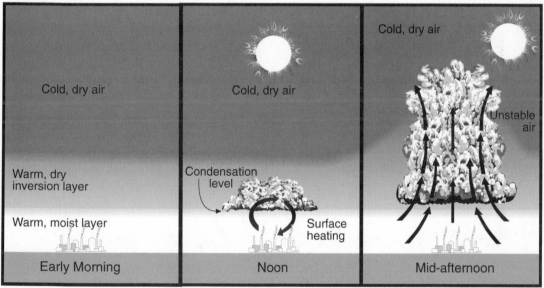

Figure 12-10. An upper-air inversion.

Data from the two upper-air stations in the region, Peoria, Illinois and Paducah, Kentucky, provide an example of typical severe weather conditions. Data for Peoria are listed in Table 12-1; Figure 12-11 on the next page shows the temperature and dew-point temperature profiles for Peoria. The data for Paducah follows.

16. Label the upper-air inversion layer on the Peoria temperature profile graph.

17. Following the Peoria example, use Figure 12-12 to plot the Paducah temperature and dew-point data at each pressure level indicated in Table 12-2.

159

TABLE 12-1

Peoria, Illinois upper-air data, June 2, 1990 7:00 PM CDT

Pressure (mb)	Height (meters)	Temp. (° C)	Dew Point (° C)	RH (%)	Wind Dir. (Deg.)	Wind Speed (Knots)
200	12148	-52.4	No Data	No Data	No Data	No Data
276	10011	-40.0	-52.8	22	224	91
300	9432	-36.1	-49.7	22	222	87
400	7368	-23.1	-38.2	23	239	62
500	5711	-10.0	-25.1	28	244	68
619	4037	0.1	-16.0	28	233	64
732	2661	8.0	-6.3	36	241	58
750	2465	5.0	2.7	85	244	52
850	1427	13.8	7.3	65	247	43
900	940	18.5	8.4	51	248	45
980	200	25.4	15.2	53	230	17

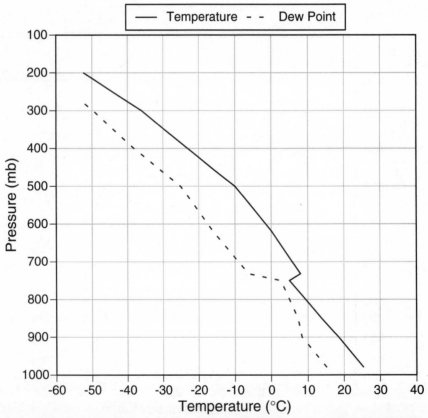

Figure 12-11. Temperature profile, Peoria, Illinois June 2, 1990 7:00 PM CDT.

160

TABLE 12-2
Paducah, Kentucky upper-air data, June 2, 1990 7:00 PM CDT

Pressure (mb)	Height (meters)	Temp. (° C)	Dew Point (° C)	RH (%)	Wind Dir. (Deg.)	Wind Speed (Knots)
200	12309	-57.6	No Data	No Data	257	56
300	9615	-36.2	-48.8	25	252	49
400	7550	-19.5	-49.5	19	260	49
500	5852	-7.6	-37.6	19	251	51
600	4408	0.9	-9.0	48	250	56
700	3145	10.5	-0.5	46	248	62
746	2611	12.3	3.4	54	248	60
765	2401	11.1	9.6	90	247	56
850	1510	17.5	13.6	78	236	47
900	1017	20.6	No Data	No Data	222	43
2996	126	28.4	22.8	71	180	17

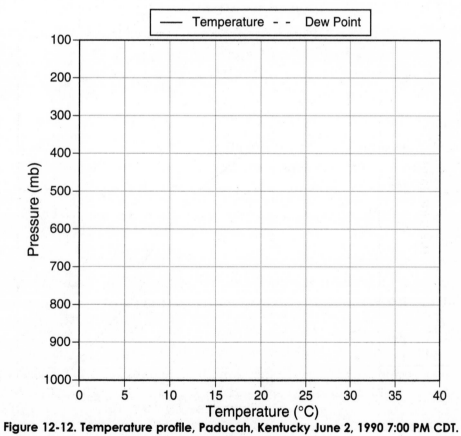

Figure 12-12. Temperature profile, Paducah, Kentucky June 2, 1990 7:00 PM CDT.

161

18. *How would you characterize the air mass below 750 mb at Paducah (e.g., warm or cool, moist or dry)?*

19. *How would you characterize the air mass above 500 mb at Paducah?*

20. *Where do you find the sharpest vertical change in temperature?*

Vertical Wind Shear

Tables 12-1 and 12-2 list the wind direction in degrees. To interpret these values, consider a circle of 360°. North is at the top of the circle (0° or 360°), east is on the right (90°), south is at the bottom of the circle (180°), and west is on the left (270°). For example, Paducah, Kentucky had a southwest wind (222°) at 1017 meters.

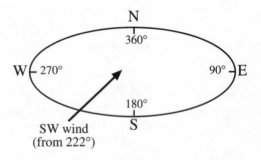

Figure 12-13

Wind speed and direction are important because the temperature and moisture characteristics of air advected into a region can determine its stability. When wind direction changes with height, advection can bring warm moist air into certain layers of the atmosphere and cool, dry air into others.

Vertical changes in wind speed and direction influence the severity of thunderstorms. Air mass thunderstorms develop when winds are light throughout the troposphere. Severe thunderstorms form when wind speed and direction change rapidly with height. Some severe thunderstorms will be multicell squall-line storms; others may develop into rotating supercell thunderstorms. The formation of supercell thunderstorms, which produce the most violent tornadoes, requires strong winds just above the surface that increase in speed and veer (turn clockwise) with altitude.

21. *Draw arrows showing Paducah wind direction at each pressure level depicted in the diagram on the next page. Make the length of your arrows proportional to wind speed.*

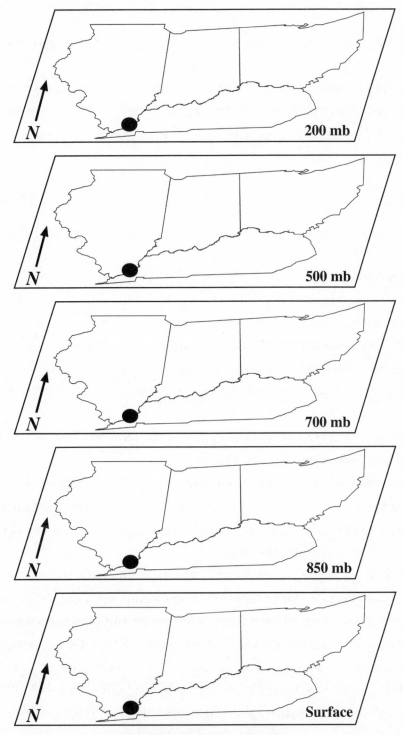

Figure 12-14. Vertical wind shear.

22. *How would vertical changes in wind direction influence the type of air masses advected at different atmospheric levels?*

23. *Why would temperature and moisture differences between the lower troposphere and the middle and upper troposphere contribute to severe weather?*

Review Questions

Describe the major processes associated with the mature stage of a thunderstorm.

What surface weather patterns lead meteorologists to predict severe thunderstorms?

What upper-air patterns would support a forecast for severe thunderstorms or tornadoes?

HURRICANES

Introduction

How do meteorologists forecast hurricanes and their potential for damage? Hurricanes are violent storms characterized by a low pressure center, high winds, heavy rain, and rough seas. They form over tropical oceans and are fueled by the evaporation of warm water and conversion of thermal energy into kinetic energy. This lab examines the structure, energy, and movement of hurricanes and uses Hurricane Hugo as a case study to illustrate these concepts.

Circulation within a Hurricane

Air circulates around and in toward a hurricane's low pressure center—the *eye* of the storm. Surface wind speeds are light in the eye, reaching a maximum approximately 25 km from the eye (an area called the *eye wall* because of its tall cumulonimbus clouds), and diminish with distance from the eye wall.

1. ***Draw the surface winds around the eye of the Northern Hemisphere hurricane depicted in Figure 13-1.***

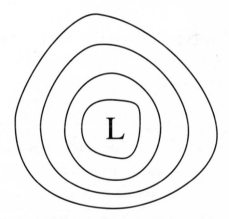

Figure 13-1. The eye of a hurricane.

Figure 13-2 uses a cross section to show the horizontal and vertical motion within hurricanes. As air moves toward the eye, the storm gains thermal energy through evaporation off the warm ocean surface. Air rises near the storm's center, forming tall cumulonimbus clouds that constitute the eye wall. At high levels of the troposphere, air diverges away from the central eye (an area of weak high pressure). This air loses energy through long-wave radiation emission, cools, and subsides at the dry adiabatic lapse rate several hundred kilometers from the eye. Some of the upper-level air sinks to the surface where it flows back toward the eye. In addition, subsidence in the eye itself reduces cloud cover.

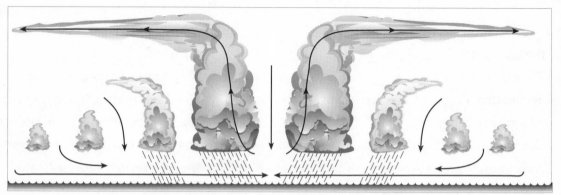

Figure 13-2. Circulation within a hurricane.

Just above the tropical waters where hurricanes form, air is uniformly warm. However, moisture content of the air varies considerably between surface air near the eye wall, where it is very high, and surface air at locations away from the storm center, where subsidence makes moisture content relatively low.

2. *How does subsidence at the storm's periphery make surface air relatively dry?*

3. *The amount of evaporation that takes place over the ocean depends on sea-surface temperature, wind speed, and the pre-existing atmospheric moisture. Consider the state of each of these variables and explain how they influence evaporation rates as air moves across the water surface from the periphery to the interior of the storm.*

4. *How are the processes described in question 3 similar to those producing lake-effect snow around the Great Lakes in late fall and early winter?*

Hurricane Hugo

Hurricane Hugo began as a cluster of thunderstorms off the West Africa coast on September 9, 1989. Over the following two weeks it developed into a strong hurricane, striking Guadeloupe, the U.S. Virgin Islands, and Puerto Rico before making landfall on the U.S. coast. The storm caused a total of 86 deaths; its 20-foot storm surge and severe winds contributed to damage exceeding $9 billion. We will examine this storm to learn about the paths of hurricanes, as well as their destructive winds and storm surge.

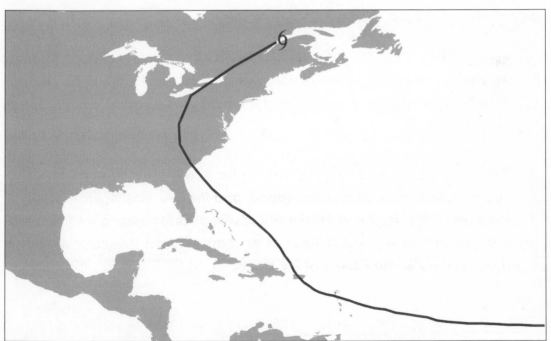

Figure 13-3. The path of Hurricane Hugo, September 1989.

167

Figure 13-4 shows the 700-mb wind speed at four locations when Hugo struck the South Carolina coast, moving toward the northwest. While the stations are equally distant from the storm's center, wind speeds vary, showing the effect of the storm's forward speed.

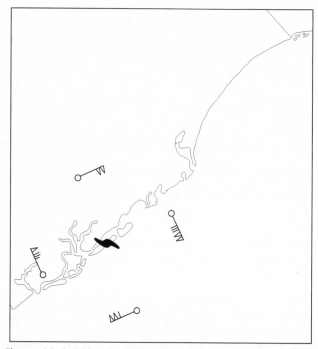

Figure 13-4. 700-mb map at landfall of Hurricane Hugo.

5. *Determine the approximate forward speed of Hugo at landfall by comparing the difference in wind speed to the right of the storm's track with the speed to the left of the storm's track.*

6. *Why is wind speed reduced in the southwest quadrant of the storm?*

168

Figure 13-5 shows Hugo's path through South Carolina. It strikes the coast on September 21 at 11:05 PM Eastern Standard Time (2305 EST) and leaves the state southwest of Charlotte, North Carolina at 5:40 AM September 22.

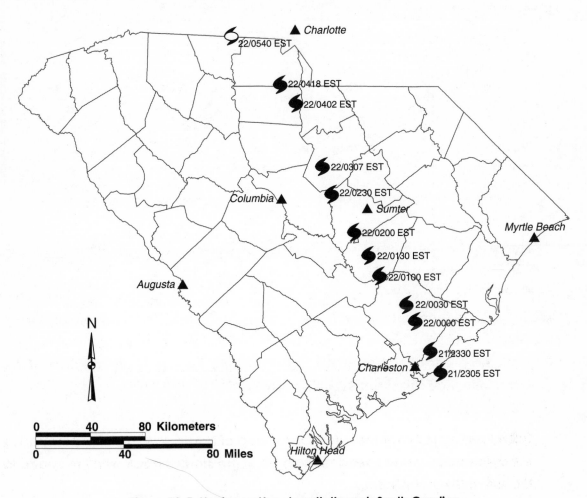

Figure 13-5. Hurricane Hugo's path through South Carolina.

7. **Locate the storm position at 2:00 AM EST on September 22. Draw arrows to indicate the wind direction in a 40-kilometer range around the storm.**

8. **Table 13-1 lists hourly pressure and wind data for Sumter, SC. When did the storm come closest to Sumter?**

9. **Explain the drastic change in wind direction at Sumter that occurs between 12:55 AM and 2:55 AM.**

TABLE 13-1

Date and Time (EST)	Pressure (mb)	Wind Direction	Wind Speed (knots)	Peak Gust (knots)
9/21 8:55 PM	993.3	010	26	34
9/21 9:55 PM	990.7	020	26	34
9/21 10:55 PM	986.5	020	30	NA
9/21 11:55 PM	979.4	020	35	47
9/22 12:55 AM	968.9	030	50	74
9/22 1:55 AM	951.6	050	58	95
9/22 2:55 AM	961.8	160	35	56
9/22 3:55 AM	976.8	160	35	51
9/22 4:55 AM	985.1	170	20	50
9/22 5:55 AM	990.7	170	24	38

Hurricanes create a *storm surge*—a dome of water as wide as 100 km—formed because of low pressure and, more important, strong winds which "push" water forward. When this dome approaches a coast, water "piles up" causing extensive coastal flooding. Shortly after Hurricane Hugo struck, the United States Geological Survey measured evidence of Hugo's storm surge and compiled the data seen in Table 13-2.

10. Mark the areas of highest storm surge in Figure 13-6. How did the path of the storm influence the storm-surge pattern?

TABLE 13-2

Average Storm-tide Elevation (feet above mean sea-level)[1]

Location	Feet	Location	Feet
Rockville	5.7	McClellanville	16.4
South Kiawah Island	10.6	Santee Point	12.1
Folly Beach	12.1	North Island	8.2
Charleston	10.9	Pawleys Island Beach	12.8
Sullivans Island	15.8	Murrells Inlet	12.6
Isle of Palms	16.2	Surfside Beach	12.9
Bull Island	16.2	Myrtle Beach	13.9
Bull's Bay	20.2	North Myrtle Beach	11.1

[1]These measurements are based on the *National Geodetic Vertical Datum* of 1929.

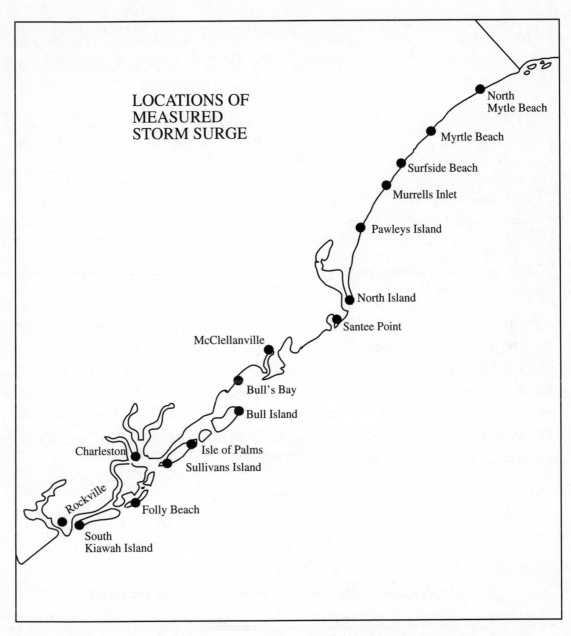

Figure 13-6. Locations of measured storm surge along the South Carolina coast.

11. *What other factors might have influenced storm-surge height along the South Carolina coast?*

General Circulation during Hugo

As powerful as a hurricane is, its path is highly dependent on continental-scale pressure and wind patterns. The 500-mb maps for September 19–22, 1989 are reproduced below to show upper-air wind flow during the week of Hugo.

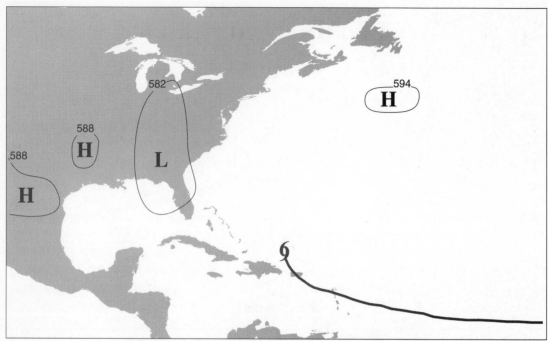

Figure 13-7. 500-mb map, September 19, 1989, 1200z.

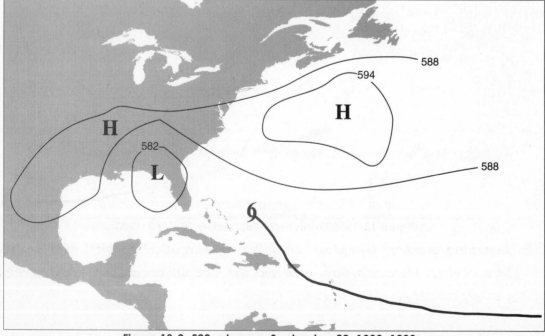

Figure 13-8. 500-mb map, September 20, 1989, 1200z.

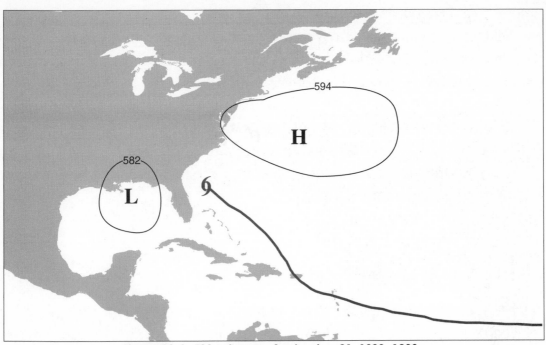

Figure 13-9. 500-mb map, September 21, 1989, 1200z.

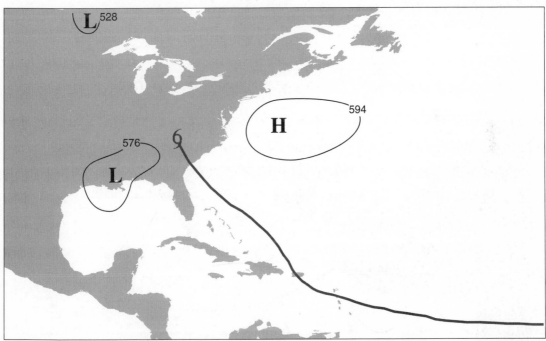

Figure 13-10. 500-mb map, September 22, 1989, 1200z.

12. *Assuming gradient wind flow, draw the winds around the high and low pressure centers for each day. How did the circulation around these pressure centers influence Hugo's path?*

This module will guide you through the formation, structure, energy, and movement of tropical cyclones with a series of sketches, radar and satellite imagery, and animations. Navigate through the module and answer the questions for each section below.

On the page titled **Formation | Sea-surface Temperatures**

13. *Briefly describe how sea-surface temperatures change between the beginning, middle, and end of the Atlantic hurricane season. What accounts for these seasonal changes?*

14. *How does this seasonal pattern of hurricane origin points change from the early season (June and early July), to mid-season (late August and early September), to the late season (late October and November) and how do these changes relate to sea-surface temperatures?*

On the page titled **Formation | Easterly Wave Animation**

15. *Describe how the size, structure, and position of the animated easterly wave changes as it moves across the Atlantic and Caribbean Sea.*

16. *Given that the animated images are separated by six-hour intervals, what is the average speed of the easterly wave moving from 47˚ W to 72˚ W (from the beginning of the animation until the end)? Note that at 15˚ N latitude, 1˚ longitude equals 108 km.*

On the page titled **Structure | Surface Conditions**

17. *Click on the red dots showing Hurricane George's (1998) path. With each click you will see the meteorological conditions recorded at buoys in the Gulf of Mexico. Choose a time segment of three consecutive dots and explain the changes in meteorological conditions that occur.*

On the page titled **Energy | Cross Section**

18. *Examine the cross-section sketch of a hurricane by rolling over each letter. At the ocean surface, how big a temperature difference exists between the eye and the outer periphery of the storm?*

19. *The thermodynamic efficiency of a hurricane depends on the temperature difference between those areas where heat is added and where it is lost. How different are temperatures between the sea-surface and the cloud tops?*

On the page titled **Energy | Landfall**

20. *Examine the animation of Hurricane George making landfall in Louisiana. How do its features change once it reaches land? Why?*

On the page titled **Path** | **Past Tracks**

21. Examine the past tracks of Atlantic hurricanes shown in the first page of the module's "Path" section. Discuss the general circulation features that influence the paths of the hurricanes shown in this graphic.

691   96366

Review Questions

What fuels a hurricane?

Under what conditions will a hurricane lose its fuel supply?

Describe wind speed and direction within a hurricane in the Southern Hemisphere.

What factors control the storm surge associated with a hurricane?

How do meteorologists forecast the path of hurricanes?

Review Questions

What aspects of climate differ between regions with rain forests and regions with savanna vegetation?

How do mid-latitude deserts differ from sub-tropical deserts?

Many of the world's great wines come from Mediterranean climates. What does this say about the climate conditions favored by the best wine-producing grapes?

The sub-arctic climate supports coniferous forests. How can a forest survive the cold temperatures and low precipitation that characterize this climate type?

What climate type would you like to live in? What aspect of the seasonal temperature and precipitation patterns appeals to you? What controls these patterns?

WORLD CLIMATES

Materials Needed
- world atlas
- climate data

Introduction

In the previous lab you examined general characteristics of Köppen's climate types. But how could you learn more about the climate of a specific place that you wanted to visit? This lab is designed to help you understand the climate of a particular place in greater detail. It should lead you to climate information available at your library, on the Internet (e.g., www.worldclimate.com), and from other sources.

Exercise

1. *Describe the climate of any location in the world.*

   a. You should begin by choosing a place and acquiring monthly temperature and precipitation data. This can be done using your textbook, climatic data references, the Internet, atlases, climate atlases, maps, or other data sources.

   b. On the next page, sketch a climograph for the city you chose.

   c. Describe the average temperature and precipitation patterns, and explain the location's climate with regard to climatic controls.

   d. Where information is available, identify and describe years that differ significantly from average conditions. You may also consider describing how vegetation patterns at your location reflect climate.

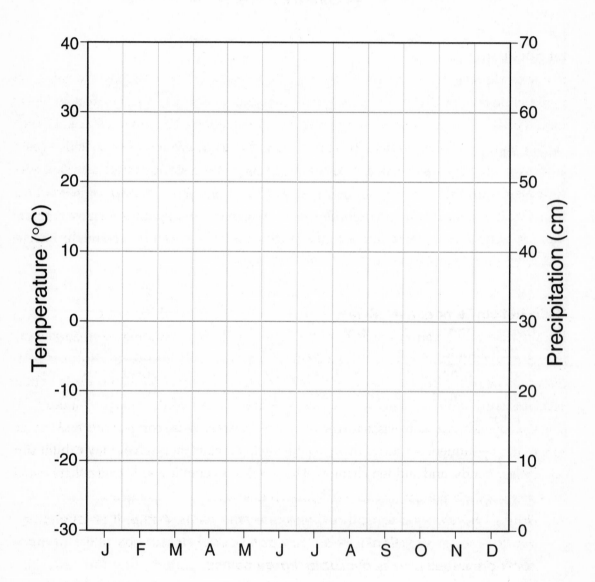

Climatic Variability and Change

Introduction

Is our climate changing? How can we tell? With recent interest in global warming, climatic variability and change have become popular topics. This interest has, however, highlighted the uncertainty involved in our understanding of the climate. It can be difficult to identify climate variability and change and to determine their specific causes. Moreover, the very way we collect data may influence the historic climate record. Added to these difficulties, climate change often involves times scales extending beyond our written climate records requiring scientists to develop *proxy* measures for past climates. This lab will focus on some of the methods climatologists use to detect climate changes and some of the problems they have with their data.

Hypothetical Temperature Curves

With statistics, we can examine a time series—i.e., how a particular climate variable changes over time. One way of examining a time series is to break out its *component parts*. Statistical analysis can reveal different aspects of the time series. There may be a regular *cycle* present (Figure 17-1), an *upward trend* (Figure 17-2), or *sudden changes* (Figure 17-3). The goal of statistical analysis is to uncover a link between these components and the factors that may influence or force them, as the forcing factors themselves may exhibit similar cycles, trends, and sudden changes. The process is complicated because there usually is a *random* component (Figure 17-4) to each time series.

1. *Figure 17-5 shows a sample temperature time series. Rather than analyzing this time series statistically, see if you can visually detect any of the components discussed previously. Label those you find.*

2. *What mechanisms could produce cyclical climate fluctuations?*

3. *What could cause a linear increase in temperature over time?*

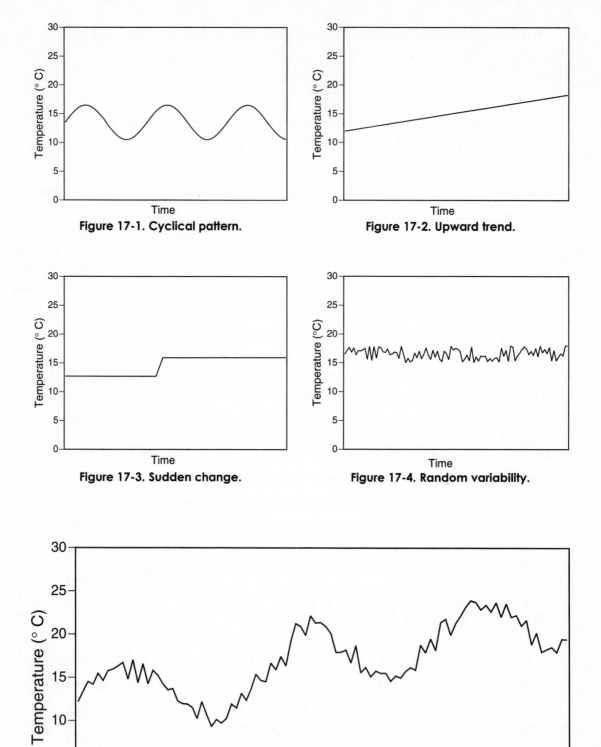

Figure 17-1. Cyclical pattern.

Figure 17-2. Upward trend.

Figure 17-3. Sudden change.

Figure 17-4. Random variability.

Figure 17-5. Temperature time series.

4. *What could cause a sudden increase or decrease in temperature?*

5. *Examine the New Haven,Connecticut,annual temperature anomaly curve (Figure 17-6). The data are plotted as °C deviations from the 1951–1970 mean. Negative values show below-normal temperatures, positive values above-normal temperatures. Comment on any patterns in the data such as cycles, gradual trends, or dramatic changes.*

6. *Which century is generally warmer, the 19th or the 20th?*

7. *The bars in Figure 17-6 show the annual dust veil index, a proxy for volcanic activity. The two largest spikes correspond to the eruptions of Mts. Tambora, Indonesia (1815) and Coseguina, Nicaragua (1835). How does the New Haven temperature record respond to those events.*

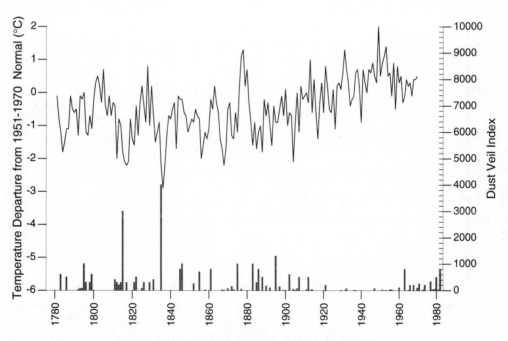

Figure 17-6. New Haven temperature variability.*

Climatic Change During the Instrumental Record

Estimates of climatic variability and change during the past 150 years are primarily derived from the *instrumental* record (i.e., compiled from direct observations with instruments). This record is more precise than indirect evidence, but is not without problems. Sites in Europe and North America are disproportionately represented in the early part of the record and perhaps skew what we know about global climate. The growth of cities during this period also creates problems with the record since urban climates differ significantly from surrounding rural areas (Table 17-1).

TABLE **17-1**

Variable	Urban Environment Compared to Rural Environment
Temperature	0.5°–1.5° C higher
Solar Radiation	15%–30% less
Precipitation	5%–15% more
Wind Speed	25% lower

8. *Why would average temperature generally be higher in the city than in surrounding rural areas?*

9. *Table 17-1 shows that regional-scale urban winds are generally lower than those in surrounding rural areas. What do you think causes this? How do low wind speeds contribute to the urban heat island? By contrast, the microscale winds around buildings can be relatively fast. Why?*

Figure 17-7 shows a 65-year record of annual mean minimum temperature for the Atlanta, Georgia international airport and Newnan, Georgia, located in a more rural setting approximately 45 km southwest of Atlanta.

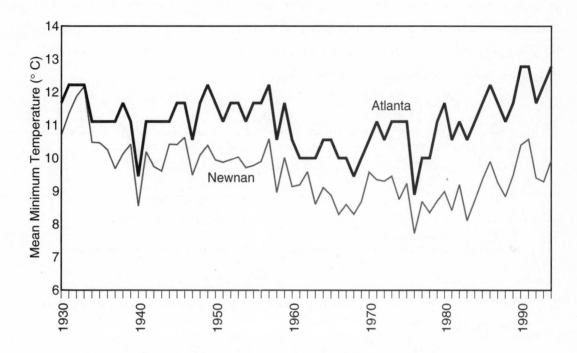

Figure 17-7. Atlanta and Newnan temperatures, 1930–1994.

10. *Using Figure 17-7, develop an argument supporting the idea that over time Atlanta's temperature has increasingly been influenced by an urban heat island.*

11. *How could urban heat islands influence our understanding of temperature changes during the past 150 years?*

Data from a study in Orlando, Florida show how a city can affect diurnal temperatures. The graph below compares the January 11th and 12th hourly temperatures at stations in Orlando's urban core with those from park locations on the outskirts of the metropolitan area.

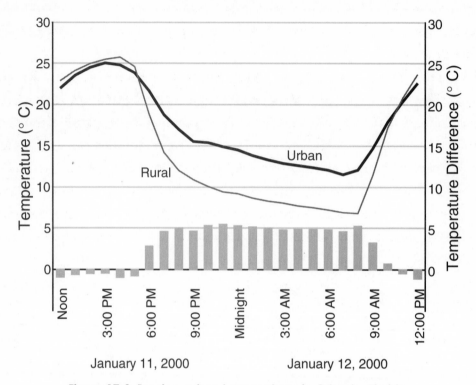

Figure 17-8. Rural vs. urban temperatures in Orlando, Florida.

12. *During what hours is the urban heat island most prominent? Why do you think it is more clearly defined during certain times of the day?*

Reconstructing Past Climate

Many of the physical mechanisms producing climate change act over long time periods and are, therefore, not evident in the relatively short *instrumental* climate record encompassing approximately the last 100 years. To uncover the details of past climate we must reconstruct the climate record using proxy data. Such reconstruction relies on the sensitivity of certain phenomena to climate. Plant and animal species, for example, respond to certain climate conditions and their thresholds and their fossil record can indicate the presence of these conditions.

Tree ring analysis has been used to reconstruct the climate over decades, centuries, and even several millennia. It assumes that annual tree growth is limited or augmented by a particular climatic variable—such as summer temperature or winter and spring precipitation. For example, a narrow annual growth ring might indicate a dry year, a wide ring a wet year. Data from an Iowa tree ring study provide an example. Two researchers, Daniel Duvick and T. J. Blasing, cored white oak trees in south-central Iowa and correlated their annual ring width with precipitation during the preceding twelve months. Notice the relationship between precipitation and annual growth for the period 1880-1980 (Figure 17-9).

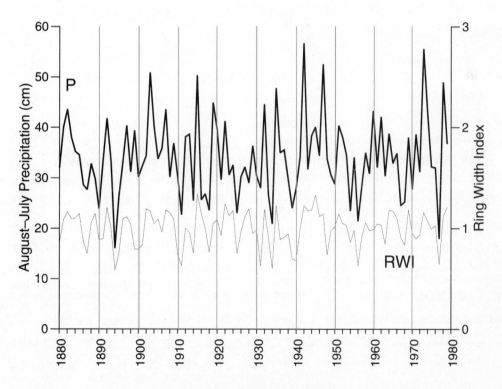

Figure 17-9. Tree-ring analysis.*

13. *Is the relationship stronger during wet years or dry years? (Hint: circle the four driest seasons and the four wettest seasons and examine the corresponding ring width. Do the driest years correspond to the most narrow rings? Do the wettest years correspond to the widest rings?)*

14. *Some climatologists have identified a 20-year drought cycle in the Central U.S., including the 1890s, 1910s, 1930s, and 1950s. How do the Iowa data support this finding?*

15. *Rather than identifying the entire decade as dry, how might you better characterize precipitation during the 1910s or 1930s?*

While tree rings provide detailed records for relatively recent climate history, many factors causing climatic change operate at longer time scales. The advance and retreat of glaciers, for example, have periods of tens of thousands of years or more. Fortunately, ocean fossils and ice cores preserve climate records at these longer time scales. Scientists have used two oxygen isotopes, ^{18}O and ^{16}O, found in ocean fossils as evidence of climatic change. Since ^{16}O evaporates more readily than ^{18}O, oceans are richer in ^{18}O during glacial advance when water moves from oceans to continental glaciers. The shells of microorganisms produced during these periods preserve the ocean's higher ^{18}O concentrations. Ice cores from Greenland and Antarctica also preserve evidence of past climate or factors that influence climate. Air bubbles trapped in the ice, for example, can reveal past concentrations of greenhouse gases.

Long-term variations in the path, tilt, and precession of the earth's orbit (Milankovitch cycles) strongly influence glacial advance and retreat. Glaciers retreat when these cycles combine to amplify the seasons in the Northern Hemisphere creating relatively warm summers and relatively cool winters. Glaciers advance when the seasons are moderated—i.e., have relatively cool summers and relatively warm winters.

Figure 17-10 shows estimates of ice volume, CO2 concentrations, June solar radiation at 60° N latitude for the past 160,000 years. Use it to answer questions 16-19.

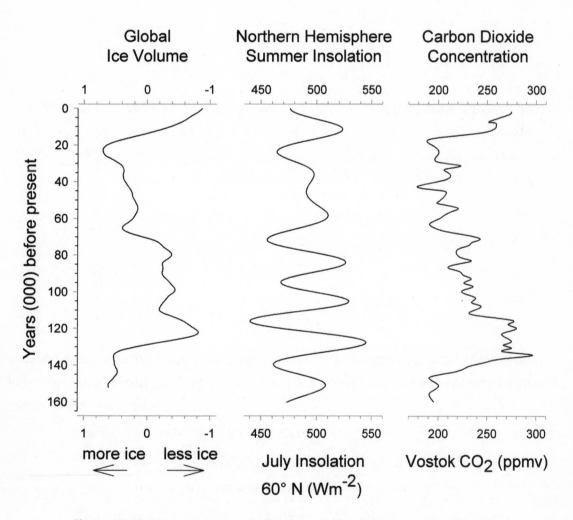

Figure 17-10. Proxy measures of climate variables for the past 160,000 years.*

16. *Examine the last glacial advance, 20,000 years ago. Was June insolation relatively high or relatively low at this time? How would a relatively cool summer contribute to glacial advance?*

17. *How could the CO_2 concentrations 20,000 years ago contribute to cooler global temperatures?*

18. *Scientists examining paleoclimate records are often curious about which variables lead and which lag. How does the timing of June insolation minima relate to glacial ice volume at 135,000, 65,000, and 20,000 years before present? Which variable leads and which lags?*

19. *How is the relationship between ice volume and CO_2 at these same times more complicated?*

Optional Exercise: The Climate Record in Your Area

Examine the instrumental climate record for your area. You may want to collect data from the National Oceanic and Atmospheric Administration (NOAA) Climatological Data Series, which is found in many government documents libraries, is also available from various Internet sources (e.g., http://cdiac.esd.ornl.gov/r3d/ushcn/ushcn.html). Does the instrumental record show any special patterns over time? Can you make any general statements about climatic variability and change for your area?

You may be interested in the local climate during times preceding the instrumental record. Perhaps there are resources in a historical library such as records of missionaries, traders, or early settlers that could help you to reconstruct the climate. Often diaries provide information about general temperature and precipitation patterns of the past.

Can you think of some other way that you could learn about the past climate of a particular area?

Review Questions

What challenges face climatologists interested in reconstructing the climate of the past?

Past climates could provide clues to future changes, but how does predicting the future climate become more complicated than simply extrapolating from past records?

What is the basis for using tree-ring analysis to reconstruct the climate? What are the potential advantages and disadvantages of this method?

---

*Data from the IGBP PAGES/World Data Center for Paleoclimatology, NOAA/NGDC Paleoclimatology Program, in Boulder, Colorado, USA and the Carbon Dioxide Information Analysis Center (CDIAC) were used in this lab.

SIMULATING CLIMATIC CHANGE

Introduction

What if you could control the earth's climate? In this lab you will conduct a series of "experiments" on the climate using the one-dimensional energy balance model developed by Professor James E. Burt at the University of Wisconsin—Madison. Such mathematical models of the earth's climate allow us to measure environmental factors such as atmospheric carbon dioxide, volcanic activity, solar output, or orbital characteristics.

Running the Energy Balance Model

Your instructor will provide basic instructions about running the model. When you start the software, you will see a simple menu showing the model's basic parameters and settings (Table 18-1). Throughout this lab you will be adjusting these settings to simulate the climatic response to specific changes.

TABLE **18-1.**

Type of Run: GLOBAL RUN		
Parameter Name	Current Setting	Change from Normal
Solar Constant	1367.0 W/m²	0.0%
Atmospheric Transmissivity	0.890	0.0%
Carbon Dioxide Concentration	350.0 ppm	0.0%
Global Surface Albedo	0.121	0.0%
Albedo Feedback	0.000 / °C	0.000 / °C
Eddy Diffusion Factor	1.0	0.0%
Ocean Mixed Layer Depth	75.0 meters	0.0%
Year	1990.0 AD	0.0 years
Eccentricity	0.017	0.0%
Day of Perihelion	2.8	0.0 Days
Obliquity	23.45°	0.00°
Run e**X**it **G**raphs **T**ables re**S**et **I**nfo		

Try moving through the menu by using the arrow keys. As you do, the window at the bottom of the screen explains the importance of each parameter. You can change a setting wherever the ">" prompt appears. After you make all appropriate changes, you run the model by typing **R** or by moving to the command **Run** at the bottom of the menu and pressing the **Enter** key.

1. *Run the model with the current settings to simulate present conditions. What is the global annual average temperature?* _____° C

You will now conduct a series of *sensitivity experiments*. That is, you will test to see how sensitive the modeled climate is to changes from present conditions.

Solar Variability

One theory of climatic change suggests that the sun is a variable star and that changes in its output could lead to climatic changes. How sensitive is temperature to changes in solar output?

2. *Decrease solar output by 1% (-1% change from normal) and run the model. How much cooler is the global temperature?* _____° C

Volcanic Activity

Volcanic activity can also affect global climate change. We can simulate the effect of volcanoes in this model by adjusting atmospheric transmissivity—the proportion of incident solar radiation that reaches the surface when the sun is directly overhead. The normal model setting is 0.89. Higher values correspond to greater transmissivity of solar radiation; lower values correspond to less transmissivity, or greater beam depletion.

3. *Reset the parameters by striking the "s" key. Then decrease the transmissivity by 1% (-1% change from normal) to simulate the effect of increased volcanic activity. What is the resulting temperature?* _____° C

4. *Explain why volcanic activity has such an effect on temperature.*

5. *Reset the model parameters by striking the "s" key.*

6. *Why would a 1% change in transmissivity produce a greater temperature change than a 1% change in the solar constant?*

Carbon Dioxide

Natural causes can alter climate, but human activity could also cause climatic change. For example, increasing concentrations of atmospheric carbon dioxide (CO_2), caused by the burning of fossil fuels, could lead to climatic change.

7. *Some projections suggest that CO_2 concentrates could reach 560 ppm by the next century. What is the model's global temperature corresponding to this change?* _____° C

8. *Explain why an increase in atmospheric CO_2 could lead to higher temperatures.*

We have examined the sensitivity of climate to changes in individual variables. Given the complexity of the earth's climate, it is possible that several changes could occur simultaneously, obscuring the effects of each individual change. Let us first examine how two independent changes, occurring together, could influence climate.

9. *Set the model for a seasonal run and increase CO_2 concentration to 560 ppm and decrease the solar constant by 1%. What is the resulting temperature?* _____° C

10. *Run the model to determine how much solar output would have to decrease in order to obscure the effect of an atmospheric CO_2 concentration increase from 350 to 560 ppm?* _____%

225

Albedo Feedback

In some cases, changes in variables that influence climate are interrelated. We will now examine how temperature changes at certain latitudes could change albedo, which in turn has an effect on the climate system. This will require a number of model parameter changes.

11. *Reset the model parameters by striking the "s" key. Then adjust the type of run to "seasonal." Next, adjust the albedo feedback to "-0.13." (This last step allows the model to calculate albedo from surface temperatures instead of using average albedo values.)*

12. *How would increasing temperature in high latitudes change albedo during winter?*

13. *Run the model and record the global temperature* _____*˚ C (The equilibrium temperature may differ slightly from your previous results because of nonlinear relationships within the model.)*

Variations in the Earth's Orbit

In an attempt to explain certain climatic changes a Yugoslavian astronomer, Milutin Milankovitch, calculated variations in three components of the earth's orbit including its *eccentricity*, *precession*, and *obliquity*. These variations affect seasonal and latitudinal receipt of solar radiation and they partly explain periods of glacial advance and retreat. The three orbital parameters can be adjusted in the model using the last three parameters of the main menu. In addition, you can change the year to explore the effect of past or future orbital variations.

14. *Retain the settings for albedo feedback (-0.13) and seasonal run type. Note the following values:*

> *day of perihelion* _____
> *eccentricity* _____
> *obliquity* _____

15. Now set the year to 25,000 years before present by typing "-25,000" in the "change from normal" column. Record the new values for:

     day of perihelion   _____

     eccentricity   _____

     obliquity   _____

In the next three questions, consider each of the orbital parameters individually:

16. *What if the perihelion date changed from January 3 to June 10 and the other orbital parameters remained at 1990 values? How would this affect the intensity of the seasons in each hemisphere?*

17. *How would a decrease in the tilt of the earth change the intensity of the seasons?*

18. *Set the model to a "Seasonal" type of run and make certain that the albedo feedback is set to -0.13/° C. Run the model for the conditions of 25,000 years before present. What is the global temperature difference between this run and the control run (see your answer to question #13)?*

19. *What about the differences in temperature at specific latitudes? Use the "Graph" command to create a graph showing the temperature difference at 55° N latitude between the control run (red dotted line) and the simulation for 25,000 years BP (solid green line). During which season do you find the greatest change? How might this change favor the presence of glacial ice at this latitude?*

20. *To simulate the conditions 25,000 years ago more realistically, we might want to lower CO_2 levels by 35%. Make this adjustment while retaining the other settings and run the model. What is the new global equilibrium temperature? _____° C. Is the greatest temperature change in the same season as it was above?*

21. Change any two variables concurrently in order to examine interactive effects and run the model again. Record the new values for the two variables that you chose.

   Variable 1 _____   Change _____

   Variable 2 _____   Change _____

22. Before you run the model, indicate how you think each change would affect the model's temperature.

23. What is the new global annual temperature? _____ ° C

24. Explain the results of your experiment.

Review Questions

How could a decrease in seasonality (i.e., warmer winters and cooler summers) lead to the advance of glacial ice on continents?

In questions 11–13 you examined the importance of albedo feedback. Provide two other examples of possible feedbacks, wherein one aspect of the climate system changes and leads to a series of other plausible changes.

DIMENSIONS AND UNITS

Materials Needed
- calculator

Systems Of Measure
To understand the features and processes of the atmosphere we must consider how meteorological variables such as energy, temperature, and pressure are measured. Meteorological variables can be expressed in terms of the following fundamental *dimensions*:

     a. Time (T)

     b. Length (L)

     c. Mass (M)

     d. Temperature (D)

Several systems have been developed to express dimensions in *units*. The most common systems are the CGS (centimeter-gram-second), SI (Le Système International), and English systems. The fundamental dimensions in each system are expressed in the following units:

Dimension	CGS	SI	English
Time (T)	second	second	second
Length (L)	centimeter	meter	foot
Mass (M)	gram	kilogram	pound
Temperature (D)	Kelvin	Kelvin	°Fahrenheit

Dealing With Units
We frequently need to work with units of different dimensions (e.g., meters, calories, seconds) in meteorological calculations. This is done by treating such units as variables. Compare, for example,

$$\frac{75x}{3y} = \frac{75}{3} \cdot \frac{x}{y} = 25\frac{x}{y}$$

to

$$\frac{75\ meters}{3\ seconds} = \frac{75}{3} \cdot \frac{meters}{seconds} = 25\ m\ s^{-1}$$

In the second equation, meters and seconds are treated as variables. The symbols 75 meters and 3 seconds represent the products of 75 times meters and 3 times seconds, respectively. The symbol meters/second represents a quotient of meters divided by seconds. We may express units appearing in the denominator with negative exponents. Thus, in the above example, our solution may be written 25 meters · second $^{-1}$ or abbreviated as 25 m s^{-1}.

Metric units are most commonly used in meteorology. However, because English units are used in the United States, it is sometimes necessary to convert between the two systems. While many sources provide conversion tables and calculators often perform conversions, there are instances where conversion between *specific* units may not be readily available. There are standard methods for such cases.

When converting units from one system to another we use the notions that anything multiplied by 1 equals itself and, conversely, anything divided by itself is equal to 1. Thus, 12 inches divided by 12 inches equals 1. Similarly, 12 inches divided by 1 foot also equals 1, as does one inch divided by 25.4 millimeters. There are occasions when multiplication by unity may be used several times in converting values. For example, we may want to express 12.5 feet in meters. Using Table A-1 as a reference, we might make this conversion in the following fashion:

1. Find a conversion between metric and English units of length from Table A-1 (1 inch = 25.4 mm).

2. Multiply by a series of relevant unity values to calculate the length in the desired units (meters) and "cancel" unwanted units:

$$12.5 \; ft \cdot \frac{25.4 \; mm}{1 \; inch} \quad \textit{(English / Metric conversion)}$$

$$12.5 \; \cancel{ft} \cdot \frac{25.4 \; mm}{1 \; \cancel{inch}} \cdot \frac{12 \; \cancel{inches}}{1 \; \cancel{ft}} \quad \textit{(cancel feet, inches)}$$

$$12.5 \; \cancel{ft} \cdot \frac{25.4 \; \cancel{mm}}{1 \; \cancel{inch}} \cdot \frac{12 \; \cancel{inches}}{1 \; \cancel{ft}} \cdot \frac{1 \; meter}{1000 \; \cancel{mm}} \quad \textit{(cancel mm)}$$

We are left with:

$$12.5 \cdot \frac{25.4}{1} \cdot \frac{1 \; meter}{1000} = \frac{3810 \; meters}{1000} = 3.81 \; meters$$

Notice that each factor we multiply by 12.5 (e.g., 25.4 mm / 1 inch, 12 inches / 1 foot, and 1 meter / 1000 mm) equals 1. In addition, all units except meters cancel since dividing feet by feet, mm by mm, or inches by inches, results in a quotient of one.

Length

1 inch	=	25.4 millimeters
1 statute mile	=	1.6 kilometers
1 nautical mile	=	1.1516 statute miles
1° latitude	=	111 kilometers

Area

1 square yard (yd^2)	=	0.8361 square meters (m^2)
1 square inch (in^2)	=	6.45 square centimeters (cm^2)
1 square meter (m^2)	=	1.196 square yards (yd^2)
1 square meter (m^2)	=	10,000 square centimeters (cm^2)
1 square centimeter (cm^2)	=	0.155 square inches (in^2)

Volume

1 milliliter (ml)	=	1 cm^3

Force

1 Newton	=	0.2248 lb

Pressure

1 Newton per m^2	=	1 Pascal		
1 mm mercury	=	133.32 Pascals		
1 pound per in^2	=	6895 Pascals		
1 bar (1000 mb)	=	100,000 Pascals	=	100 kiloPascals (kPa)

Standard sea-level pressure, or 1 atmosphere, can be expressed by any of the following units:

1 atmosphere = 101325 Pascals = 1.01325 bars = **1013.25 millibars**

760 mm mercury = **29.92 inches mercury** = 14.7 lb/in^{-2}

Energy

1 Joule	=	0.2389 calories		
1 **British Thermal Unit (BTU)**	=	1055 Joules	=	252 calories
1 kilowatt hour	=	3,600,000 Joules		

Temperature

° Celsius	=	5/9 • (° Fahrenheit – 32.0)
° Celsius	=	Kelvins – 273.15

Variables Important To Meteorology

Length

Dimension: L (Length)

1. **Convert the typical altitudes of:**

    a. geostationary satellites (36,000 kilometers) = _____ miles

    b. cruising jet airplanes (37,000 feet) = _____ meters

    c. Most of earth's weather results from processes occurring below this cruising height. What is the ratio of this height to the earth's radius (6400 km)?

2. **Evangelista Torricelli's barometer showed that atmospheric pressure could push mercury up a tube (Figure A-1). "Standard" atmospheric pressure is 760 mm of mercury. This is equal to _____ inches of mercury.**

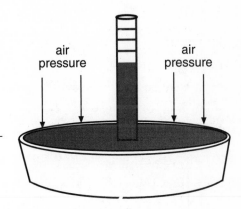

Figure A-1. Torricelli's barometer.

Area

Dimensions: L² (Length • Length)

3. **Convert the following:**

    10 in² = _____ cm²

    20 m² = _____ ft²

    (Hint: If you imagine 20 m² as a perfect square, you can determine the length of one side of that square by simply calculating the square root of the area.)

Volume

Dimensions: L^3 (Length • Length • Length)

4. What is the volume of a box with the follow-ing dimensions:

  width: 2 meters
  height: 2 meters
  depth: 2 meters

  ____8____ m^3 = _____ ft^3

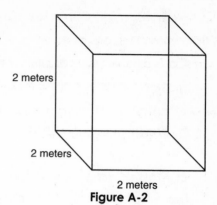

2 meters

2 meters

2 meters

Figure A-2

Density

Dimensions: ML^{-3} (Mass • Length^{-3})

Density is equal to mass divided by volume. In SI units, it is most often expressed as $kg\ m^{-3}$. Air density at sea level generally will vary between 1.1 and 1.3 $kg\ m^{-3}$, depending on air temperature and pressure.

5. If the box in Figure A-2 had a sea-level air density of 1.23 kg m^{-3}, how much would the air within the box weigh? _____ kg

Velocity

Dimensions: LT^{-1} (Length • Time^{-1})

6. If a strong west wind blows at 35 nautical miles per hour (35 knots), how fast is the wind moving:

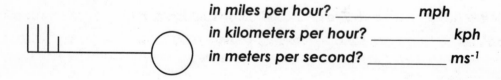

  in miles per hour? _____ mph
  in kilometers per hour? _____ kph
  in meters per second? _____ ms^{-1}

Figure A-3

Acceleration

Dimensions: LT^{-2} (Length • $Time^{-2}$)

Acceleration is defined as a change in *speed* or *direction* with time. In SI units, acceleration is expressed as meters per second squared (ms^{-2}). There are several types of acceleration important in meteorology, including gravitational and Coriolis acceleration. Acceleration due to gravity equals 9.8 ms^{-2}. That is, every second the velocity of a freely falling object increases by 9.8 ms^{-1}.

Force

Dimensions: MLT^{-2} (Mass • Length • $Time^{-2}$)

Atmospheric forces cause vertical and horizontal movement of air. In SI units, force is expressed in *Newtons* (kg m s^{-2})—the force necessary to accelerate a one-kilogram mass one meter per second every second. Newton's Second Law expresses force as:

$$Force = Mass • Acceleration$$

7. **What is the gravitational force on the <u>air</u> in the box described in Figure A-2?**
   _____ kg m s^{-2} = _____ Newtons (N)

Pressure

Dimensions: $L^{-1}MT^{-2}$ ($Length^{-1}$ • Mass • $Time^{-2}$)

Pressure is an important variable in meteorology and is best understood as force exerted over a unit area. Pressure is expressed in SI units as Newtons per meter squared (N m^{-2}), or Pascals (1 N m^{-2} = 1 Pascal).

8. **If the box in Figure A-2 rests on the earth's surface, it occupies an area of 4 m^2. What is the pressure exerted as gravity pulls on the air in the box?**
   _____ Newtons m^{-2} = _____ Pa

Another unit of pressure is the bar. It was chosen to approximate the pressure of the atmosphere at sea level. Standard atmospheric pressure equals 101,325 Pascals (Pa) and 1 bar = 100,000 Pa = 100 kPa. Pressure is commonly expressed on weather maps in millibars, i.e., one thousandth of a bar (1 millibar = 0.001 bars).

9. **Convert the following: 1 millibar (mb) = _____ Pascals (Pa)**

10. **Express standard sea-level pressure in millibars. _____ mb**
    And in kilopascals _____ kPa,

11. **Convert the following: 1024 mb = _____ inches of mercury.**

Energy

Dimensions: $L^2 M T^{-2}$ (Length^2 • Mass • Time^{-2})

Energy measures the ability to do work and thus can be considered as the measure of a force acting over some distance. In the SI system, energy is expressed in *joules*. One joule is equal to one Newton acting over a distance of one meter (1 joule = 1 N m). Another commonly used unit of energy is the *calorie*, which is derived from the CGS system. It is common to examine energy over some area (e.g., joules per square meter or calories per square centimeter). In the CGS system, one calorie per square centimeter is equal to one *langley*. Because of the variety of units used to measure energy, it is often necessary to convert between systems. Consider, for example, conversion between langleys and joules per square meter. Substituting 1 calorie per cm^2 for 1 langley we have:

$$\frac{1 \; cal}{cm^2} \cdot \frac{1 \; Joule}{0.2389 \; cal} \cdot \frac{10,000 \; cm^2}{1 \; m^2} = 41,858.5 \; J \; m^{-2}$$

Power

Dimensions: $L^2 M T^{-3}$ (Length^2 • Mass • Time^{-3})

Power is an expression of energy per unit time. In the SI system, 1 joule per second = 1 *watt*.

Like energy, power is commonly expressed per unit area. The *solar constant* is an example of this measure: it is the rate of solar energy receipt on a hypothetical two-dimensional disk, perpendicular to the sun's rays, and ignoring the effects of the atmosphere (Figure A-3). The solar constant is approximately 1.96 cal/cm^2 (langleys) per minute.

12. Express the solar constant in watts per square meter. _____

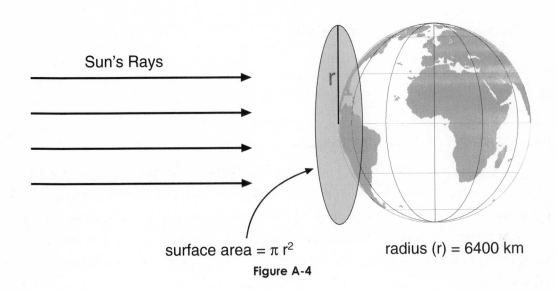

Sun's Rays

r

surface area = π r²

radius (r) = 6400 km

Figure A-4

Temperature

13. Select the maximum and minimum temperatures from a recent day and convert to ° C and Kelvins.

	° F	° C	K
Maximum Temperature	89°	____	____
Minimum Temperature	67°	____	____

EARTH MEASURES

Materials Needed
- Goode's (or other) World Atlas

Latitude and Longitude

Because of its near-spherical shape, it is convenient to measure locations on the earth in degrees. A complete circumference around the earth is equal to 360°. *Latitude* measures position relative to the widest bulge of the earth, the *equator* (0° latitude), and is expressed in degrees North or South (° N, ° S).

Since there are 90 degrees in a quarter of a sphere (360° ÷ 4), the North and South Poles are located at 90° N and 90° S latitude, respectively.

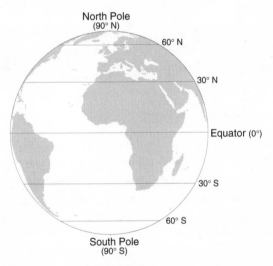

Figure B-1

1. *Find the latitude of the following locations:*

    _____ *Equator*   _____ *Kansas City, MO, USA*

    _____ *Arctic Circle*   _____ *Montreal, Canada*

    _____ *Antarctic Circle*   _____ *Tropic of Cancer*

    _____ *Mexico City*   _____ *Tropic of Capricorn*

    _____ *Bombay, India*   _____ *Cape Town, S. Africa*

Longitude measures position east or west of the *Prime Meridian*—a line passing from the North to South Pole, through Greenwich, England. Lines of longitude to the east of the Prime Meridian are expressed in degrees east (° E), and those to the west of the Prime Meridian in degrees west (° W). Halfway around the earth from Greenwich the lines of east and west longitude meet at 180°.

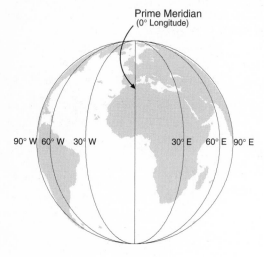

Figure B-2

Both longitude and latitude can be broken down into 60th fractions, or minutes. Minutes are broken down into 60 seconds.

1°	=	60 minutes (60′)
1′	=	60 seconds (60″)

2. **Find the latitude and longitude of:**

_____ **Los Angeles, CA, USA**   _____ **Lima, Peru**

_____ **Baghdad, Iraq**   _____ **Moscow, Russia**

_____ **Melbourne, Australia**   _____ **Rome, Italy**

_____ **Nairobi, Kenya**   _____ **Atlanta, GA, USA**

3. **The earth makes a complete rotation on its axis every 24 hours. Since the earth is 360° in circumference, by how many degrees longitude does it rotate every hour? _____. How does this relate to time zones on the earth.**

4. **Pick two places on the earth, indicate their longitude, and determine the approximate longitude and time difference between the two locations.**

LOCATION	**LONGITUDE**	**LONGITUDE DIFFERENCE**	**TIME DIFFERENCE**
_____	_____		
_____	_____	_____	_____

General Geography

Atmospheric processes are interconnected. What happens in one part of the world can affect the weather and climate in another part. Since we will make occasional reference to places around the world, it is essential that you review a bit of world geography. Study your atlas and locate the world's continents and oceans.

5. Mark the following places on the world map provided.

Continents

- Africa
- Antarctica
- Asia
- Australia
- Europe
- North America
- South America

Oceans

- Arctic Ocean
- Atlantic Ocean
- Indian Ocean
- Pacific Ocean

Mountain Ranges

- Alps
- Andes
- Appalachians
- Cascades
- Caucasus
- Himalayas
- Rockies
- Urals

Deserts

- Arabian
- Atacama
- Gobi
- Great Victorian
- Kalahari
- Mojave
- Sahara
- Turkestan Deserts (Peski Karakumy, Peski Kyzyl Kum)

Ocean Currents

- Benguela Current
- Brazil Current
- California Current
- Equatorial Counter Current
- Kuroshio
- North Atlantic Current
- Peru Current
- Southwest Monsoon Current
- South Equatorial Current

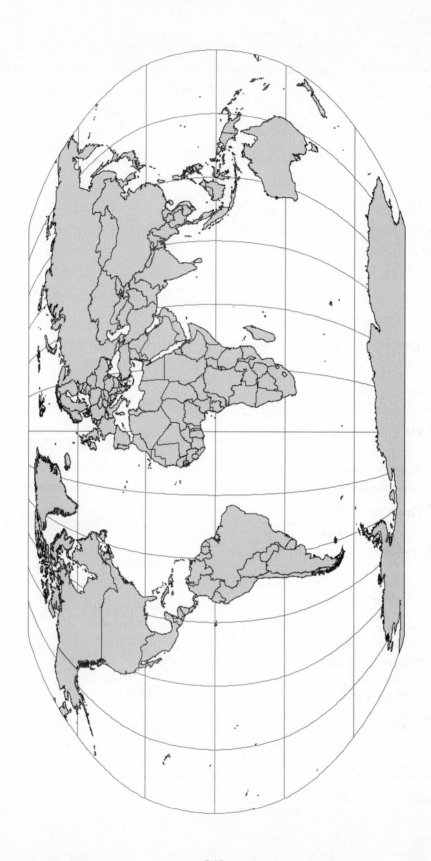

242

WEATHER SYMBOLS

Following is a short list of weather symbols you may find helpful in completing some of the labs. The National Weather Service web site (http://weather.noaa.gov/graphics/chartref.gif) has a more complete list of the weather symbols. (The appendix of your textbook may also have a fuller listing—check there as well.)

Current Weather	Barometric Tendency
● Intermittent rain	Rising, then falling
●● Continuous rain	Rising, then steady
✳✳ Continuous snow	Rising steadily, or unsteadily
ᕚᕚ Continuous drizzle	Falling or steady, then rising
☰ Fog	Steady
⛈ Thunder heard	Falling, then rising
⛈ Thunder with intermittent rain	Falling, then steady
	Falling steadily, or unsteadily
	Steady or rising, then falling

Winds	mph
	Calm
	1-2
	3-8
	9-14
	15-20
	21-25
	26-31
	32-37
	38-43
	44-49
	50-54
	55-60
	61-66
	67-71
	72-77
	78-83
	84-89
	119-123

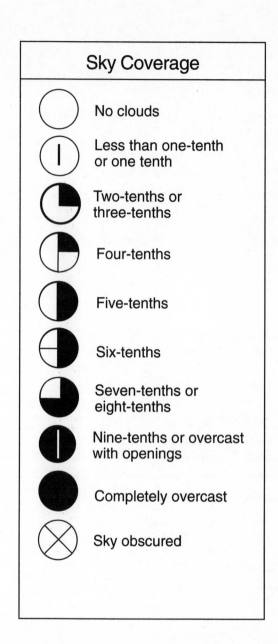

Sky Coverage

	No clouds
	Less than one-tenth or one tenth
	Two-tenths or three-tenths
	Four-tenths
	Five-tenths
	Six-tenths
	Seven-tenths or eight-tenths
	Nine-tenths or overcast with openings
	Completely overcast
	Sky obscured